Advanced Structural Ceramics

MATERIALS RESEARCH SOCIETY SYMPOSIA PROCEEDINGS

ISSN 0272 - 9172

MATERIALS RESEARCH SOCIETY SYMPOSIA PROCEEDINGS

MATERIALS RESEARCH SOCIETY SYMPOSIA PROCEEDINGS

MATERIALS RESEARCH SOCIETY SYMPOSIA PROCEEDINGS

MATERIALS RESEARCH SOCIETY CONFERENCE PROCEEDINGS

VLSI-I—Tungsten and Other Refractory Metals for VLSI Applications, R. S. Blewer, 1986; ISSN: 0886-7860; ISBN: 0-931837-32-4

VLSI-II—Tungsten and Other Refractory Metals for VLSI Applications II , E.K. Broadbent, 1987; ISSN: 0886-7860; ISBN: 0-931837-66-9

TMC—Ternary and Multinary Compounds, S. Deb, A. Zunger, 1987; ISBN:0-931837-57-x

MATERIALS RESEARCH SOCIETY SYMPOSIA PROCEEDINGS   VOLUME 78

Advanced Structural Ceramics

Symposium held December 1-3,1986, Boston, Massachusetts, U.S.A.

EDITORS:

P. F. Becher
Oak Ridge National Laboratory, Oak Ridge, Tennessee, U.S.A.

M. V. Swain
CSIRO, Victoria, Australia

S. Sōmiya
Tokyo Institute of Technology, Yokohama, Japan

MⓇS   MATERIALS RESEARCH SOCIETY
Pittsburgh, Pennsylvania

This book has been registered with Copyright Clearance Center, Inc. For further
information, please contact the Copyright Clearance Center, Salem, Massachusetts.

Published by:

Materials Research Society
9800 McKnight Road, Suite 327
Pittsburgh, Pennsylvania 15237
Telephone (412) 367-3003

Library of Congress Cataloging in Publication Data

Printed in the United States of America

Manufactured by Publishers Choice Book Mfg. Co.
Mars, Pennsylvania 16046

Contents

PART III - MECHANICAL PROPERTIES AND MICROSTRUCTURES
              OF ZIRCONIA TOUGHENED CERAMICS

PART IV - MECHANICAL BEHAVIOR OF REINFORCED CERAMIC COMPOSITES

PART V - FRACTURE AND DEFORMATION BEHAVIOR IN CERAMIC COMPOSITES

Preface

The papers contained in this proceedings were presented at
the symposium on "Advanced Structural Ceramics" held in Boston,
Massachusetts, December 1—3, and 1986 sponsored by the Materials
Research Society. The symposium addressed recent research in the
field of toughened ceramics and included that on transformation
toughening and fiber and whisker reinforced ceramics. The symposium
was international in character which is indicative of the wide interest
in exploring mechanisms to enhance the fracture resistance of ceramics.
The research presented at the symposium provides considerable insight
into our progress in these areas.

The success of any symposium also reflects the support of others.
This symposium was made possible by the financial support of the U.S.
Department of Energy through the Division of Materials Sciences, Office
of Basic Energy Sciences and the Energy Conversion and Utilization
Technologies Materials Program, Office of Renewable Energy. We would
like to especially acknowledge A.H. Heuer, D.R. Clarke, F.F. Lange,
R.J. Gottschall, and R.M. Cannon for subchairing this session and
coordinating manuscript reviews. Their efforts contributed greatly
to the success of this symposium. Finally, sincere thanks are extended
to Fauna Stooksbury who provided not only secretarial services but also
handled and coordinated many of the details required to conduct the
symposium and publish the proceedings.

P.F. Becher
M.V. Swain
S. Sōmiya

PART I

Transformation Analysis

DISPLACIVE PHASE TRANSFORMATIONS IN ZIRCONIA-BASED CERAMICS

B.C. MUDDLE
Department of Materials Engineering, Monash University, Clayton, Victoria,
Australia 3168

ABSTRACT

A review is presented of experimental observations of the mechanism and crystallography of the martensitic tetragonal to monoclinic transformation occurring both in dispersed tetragonal ZrO_2 particles in partially stabilized zirconia and in polycrystalline tetragonal zirconia. Preliminary results of determination of the orientation relationship and habit plane for the stress-activated transformation in a CeO_2-stabilized TZP ceramic are reported and compared with predictions of the crystallographic theory for the transformation. This orientation relationship is such that $(100)_m//(100)_t$ and $[001]_m//[001]_t$, and for this variant of the orientation relationship the habit plane is approximately $(301)_t$. These results are in good agreement with theoretical predictions. Progress in the application of the formal theory of martensitic transformations to the transformations in both types of system is examined critically and implications for theories of transformation toughening are discussed.

Attention is also given to tetragonal $\rightleftharpoons$ orthorhombic and orthorhombic $\rightleftharpoons$ monoclinic transformations occurring in ZrO_2 particles in thin foil specimens of partially stabilized zirconia. Formation of a metastable orthorhombic phase appears a possible, but not essential, intermediate stage in the tetragonal to monoclinic transition. However, present evidence strongly suggests that the orthorhombic structure only occurs in those particles experiencing the relaxed matrix constraints typical of thin foil specimens.

1. INTRODUCTION

Transformation-toughened ceramics constitute a new and important class of materials combining high strength and useful toughness [1,2]. Theories of transformation toughening [3] and transformation-induced plasticity [4] in zirconia-based ceramics involve consideration of the interaction between the stress field of a propagating crack and the strains accompanying the martensitic tetragonal (t) to monoclinic (m) transformation occurring within tetragonal zirconia (t-ZrO_2) in the vicinity of the crack tip. To assess the toughening that may be achieved these theories require a reliable determination of these strains and, since the transformation is displacive, this in turn requires a detailed understanding of the crystallography of the transition. For this reason the transformation has received considerable attention in pure bulk zirconia [5-11], in t-ZrO_2 particles dispersed in a ceramic matrix [11-17] and, more recently, in zirconia-based systems with a microstructure comprising polycrystalline tetragonal phase [11,18]. It is the purpose of the present paper to review briefly experimental observations relating to the t → m transformation both in dispersed t-ZrO_2 particles and in tetragonal zirconia polycrystals (TZP), and to assess progress in the application of the formal theory of martensitic transformations.

Attention will also be given to the occurrence of tetragonal $\rightleftharpoons$ orthorhombic and orthorhombic $\rightleftharpoons$ monoclinic transformations in small ZrO_2 particles which are either in unconstrained form or dispersed in the cubic

4

matrix phase in thin foils of partially stabilized zirconia (PSZ). Identification of a novel orthorhombic (o) phase in MgO-PSZ [19-21] has led to considerable speculation regarding not only its contribution to determining mechanical properties [20,22], but also its possible role in the important t → m transition. It has been suggested [19,20], for example, that the orthorhombic phase is an essential intermediate reaction product in the t → m transformation and it is thus important that the role of the orthorhombic phase be clearly established.

2.   CRYSTALLOGRAPHIC THEORY

The structural change that accompanies a martensitic transformation is characterized by the maintenance of a unique lattice correspondence between unit cells of the parent and product lattices. The existence of this correspondence implies that the change in structure may be accomplished by atomic displacements equivalent to a homogeneous deformation of the parent lattice. When combined with a rigid body rotation, R, the homogeneous lattice strain, B, implied by the correspondence defines the total lattice strain, S_t, that will generate the product lattice in its observed orientation relationship with the parent lattice [23]; i.e.

$$S_t = RB. \tag{1}$$

The total strain provides, however, an incomplete description of the transformation for it is, in most cases, incompatible with the homogeneous shape deformation that accompanies formation of the transformed volume. The shape strain, P, approximates to an invariant plane strain in which the interface plane (i.e. the habit plane) remains invariant and it is rare that the total lattice strain generates a matching plane between parent and product lattices. This apparent incompatibility may be reconciled if it is assumed that the total strain is only locally homogeneous and occurs inhomogeneously on a macroscopic scale in order that the habit plane remain an invariant plane of the shape strain. The shape strain may thus be considered the product of the total lattice strain and an additional strain which periodically relieves the accumulating misfit across the transformation interface and maintains a macroscopically undistorted plane of

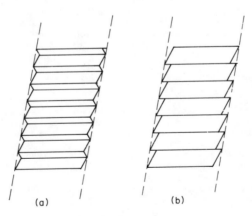

(a)   (b)

Figure 1   Schematic representation of martensite plates in which the lattice invariant strain involves (a) twinning, and (b) slip.

contact; i.e.

$$P = S_t L = RBL. \tag{2}$$

The strain L is lattice invariant and must constitute either a slip or twinning shear of the product lattice. This description of the transformation represents the basis of the formal crystallographic theory [23,24]. It implies transformation products such as those depicted schematically in Fig. 1.

In its simplest form, the theory assumes the lattice invariant strain (LIS) be a simple shear and the total lattice strain is thus the resultant of consecutive invariant plane strains,

$$S_t = PL^{-1}. \tag{3}$$

Such a strain is an invariant line strain, characterized by the existence of an invariant line, **x**, defined by the intersection of the two invariant planes, and a plane with invariant normal, **n'**, containing the two displacement directions. The strain B extends all vectors to their final lengths, leaving unrotated a set of principal axes, and is thus specified by the measured lattice parameters of the initial and final lattices and the assumed correspondence between them. The direction **x** and normal **n'** are not

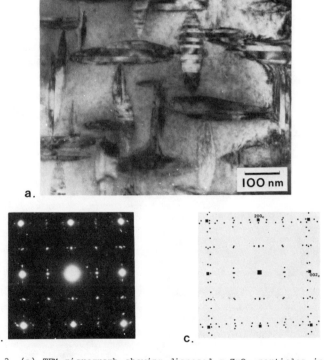

Figure 2   (a) TEM micrograph showing disposed m-ZrO$_2$ particles in MgO-PSZ, (b) corresponding $\langle 100 \rangle_c$ SAED pattern, and (c) schematic pattern expected of 24 permitted monoclinic orientations [17].

rotated by S_t, so that R is the unique rotation which reverses the rotations of $\mathbf{x}$ and $\mathbf{n'}$ due to B. If the strain S_t is specified in this manner and the plane, $\mathbf{p_2'}$ and direction, $\mathbf{d_2}$, of the LIS are assumed, then the invariant line strain may be factored uniquely into its component invariant plane strains and explicit predictions obtained for the elements of the shape strain and the orientation relationship. These predictions for the t $\rightarrow$ m transformation in metastable t-ZrO$_2$ are examined in Section 3.2

It is to be noted that for solutions to exist with a particular lattice correspondence and LIS system, the LIS plane must intersect what is termed the initial Bain cone [23], defined by those directions in the initial lattice left unchanged in length by the lattice deformation B. The vectors defined by the points of intersection remain unchanged in length as a result of both the LIS and the strain B and, depending on the rotation R, at least one of them will define an invariant line in the habit plane. The habit plane normal will thus lie in one of the planes whose poles are the intersections of the LIS plane with the Bain cone. The location of the habit plane normal within a given plane is determined by the direction of the LIS [16].

3. THE TETRAGONAL-MONOCLINIC TRANSFORMATION IN PSZ

3.1 Experimental Observations [17]

Dispersed, coherent particles of metastable t-ZrO$_2$ in MgO-PSZ take the form of lenticular plates parallel to $\{100\}_c$ planes of the cubic (fluorite) matrix phase. The orientation relationship between cubic and tetragonal lattices is such that the principal axes of the tetragonal and fluorite unit cells are parallel and, for each of three variants observed in a given matrix orientation, the c_t axis is perpendicular to the habit plane of the plate [25]. In both particles transformed athermally and those transformed under stress, the product of the t $\rightarrow$ m transition comprises parallel domains of the monoclinic phase extending either parallel or perpendicular to the original particle habit plane. The typical microstructure is shown in Fig. 2, along with the associated selected area electron diffraction (SAED) pattern.

Figure 3 shows an example of a large particle in which the monoclinic domains extend parallel to the particle habit plane and the corresponding electron microdiffraction pattern from adjacent monoclinic domains. In the schematic solution provided, the direction of the incident beam is defined parallel to $[010]_c$ and is in turn parallel to $[010]_m$ and $[0\bar{1}0]_m$ zone axes of the monoclinic domains. The trace of the domain boundaries is parallel to the trace of $(001)_m$ and the patterns for adjacent domains are related by reflection in this plane. Within experimental accuracy, adjacent domains are thus twin related, the apparent twinning plane being $(001)_m$. If an identity relationship is assumed between the principal axes of the cubic and tetragonal unit cells, then the orientation relationship between the monoclinic and tetragonal cells is such that $(001)_m//(001)_t$ and $[100]_m//[100]_t$.

For the monoclinic particle containing transverse domains in Fig. 4, the incident beam is defined parallel to $[0\bar{1}0]_c$ and is in turn parallel to $[0\bar{1}0]_m$ and $[010]_m$ zone axes of adjacent monoclinic domains. In this case, however, the trace of the domain boundary is parallel to the trace of $(100)_m$ and the domains appear twin-related about $(100)_m$. The orientation relationship implied between tetragonal and monoclinic lattices is in this case such that $(100)_m//(100)_t$ and $[001]_m//[001]_t$. This relationship differs from that observed for $(001)_m$ twin-related domains by a rotation of ~9° about the [010] axis common to the tetragonal and monoclinic lattices

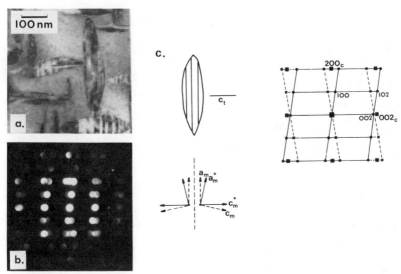

Figure 3 (a) m-ZrO$_2$ particle comprising domains parallel to (001)$_m$, (b) corresponding electron microdiffraction pattern, and (c) schematic solution to (b).

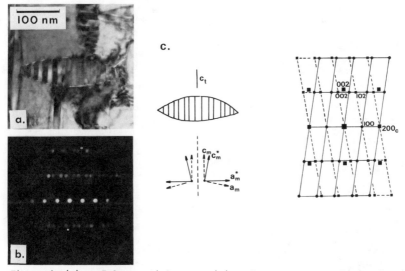

Figure 4 (a) m-ZrO$_2$ particle comprising transverse monoclinic domains parallel to (100)$_m$, (b) corresponding microdiffraction pattern, and (c) schematic solution to (b).

[17]. However, the lattice correspondence implied by both orientation relationships is such that the c_m axis is parallel to c_t.

Since there is 4-fold symmetry about the c_t axis, there are four crystallographically equivalent variants of each of the above orientation relationships and for each relationship these variants occur in two twin-related pairs. Given that there are three variants of the tetragonal lattice, two possible orientation relationships between tetragonal and monoclinic lattices, and four possible variants of each relationship, a total of twenty four different monoclinic orientations (occurring in twin-related pairs) is possible within particles observed in a given orientation of the matrix phase. The schematic $<100>_c$ diffraction pattern formed by superimposing single crystal patterns expected for all 24 monoclinic orientations is shown in Fig. 2(c). It is noted that it is in excellent agreement with that observed experimentally, Fig. 2(b).

3.2 Application of the Crystallographic Theory

In the absence of clear experimental evidence, it has been common to identify three possible lattice correspondences that may arise between t and m lattices, depending on which monoclinic axis a_m, b_m or c_m is parallel to the tetragonal c_t axis. These are commonly referred to as lattice correspondences A, B and C respectively. For application of the theory to the transformation in dispersed t-ZrO_2 particles, it is clear that the appropriate choice is correspondence C for both particle forms observed [17]. The immediate problem involves the choice of the LIS system and, bearing in mind the anticipated substructure of a martensitic product (Fig. 1), it is initially tempting to identify the twinning within transformed particles as evidence of the LIS. However, it has recently been shown [16] that neither of the twinning planes $(100)_m$ or $(001)_m$ intersect the Bain cone, defined by initial lattice directions left unchanged in length by the lattice deformation B. As indicated in Section 2, this means that neither plane, as the plane of LIS, will give rise to real solutions in theoretical calculations. Furthermore, neither of these planes is itself a potential habit plane with any choice of simple shear as the LIS [16]. If the theory is to be applicable, an alternative choice of LIS is necessary and each individual monoclinic domain within a transformed particle is to be regarded as a discrete variant of the transformation product. The junction plane between adjacent monoclinic variants is not to be confused with the habit plane for the tetragonal-monoclinic transformation.

In the most comprehensive calculations yet undertaken, Kelly and Ball [16] have recently reported predictions of the crystallographic theory assuming the LIS to be a simple shear. They have identified a number of possible shear planes giving rise to plausible real solutions and permitted all possible slip directions within each of these planes. They have eliminated those solutions involving unrealistically large values for the magnitudes m_1 and m_2 of the shape strain and LIS respectively and identified as most plausible those solutions involving the smallest values of these strains in combination. More importantly, they have recognised that for the chosen correspondence C there are four crystallographically equivalent variants of the correspondence associated with the 4-fold symmetry about the c_t axis. For certain LIS and certain variants of the correspondence taken in pairs, it is possible to generate equivalent solutions to the theory, which are at least approximately twin-related and for which the potential contact plane between twin-related variants is either $(100)_m$ or $(001)_m$. Two of four such solutions examined in detail by Kelly and Ball are reproduced in Table I; the notation 6A and 14B is that adopted in the original paper [16].

For system 6A, the solution shown has been calculated assuming that the lattice correspondence is of the form:

$$\mathstrut_m C_t \quad = \quad \begin{vmatrix} 1 & 0 & 0 \\ 0 & 1 & 0 \\ 0 & 0 & 1 \end{vmatrix} \tag{4}$$

If alternatively, an equivalent correspondence related by a rotation of 180° about $[001]_t$ is assumed, an equivalent solution is achieved in which the monoclinic variants are approximately twin related. As shown in Table I, the junction plane (or contact plane) anticipated between these two variants is exactly parallel to $(100)_t$ and only ~ 0.1° from $(100)_m$. The predicted morphology for a ZrO_2 particle transforming to a combination of such variants is shown in Fig. 5(a) and is in good agreement with the observed morphology, Fig. 4, for particles comprising transverse domains of the monoclinic structure twin-related about $(100)_m$. For system 14B, twin-related variants may be generated in a similar fashion, with a junction plane parallel to $(001)_t$ and just 0.1° from $(001)_m$. In this case the predicted morphology accounts well, Fig. 5(b), for those particles observed to comprise monoclinic variants twin-related about $(001)_m$. Figure 5 contains schematic representations of partially transformed particles to emphasize that the transformed particle comprises discrete variants of the monoclinic phase and that a distinction is to be made between the habit plane between tetragonal and monoclinic phases and the junction plane between monoclinic variants. In a fully transformed particle, no trace of

Table I Selected Predictions of the Crystallographic Theory for the Tetragonal-Monoclinic Transformation in MgO-PSZ [16]

Solution	6A	14B
Lattice Invariant Shear	$(011)[0\bar{1}1]$	$(110)[1\bar{1}0]$
m_2	0.0377	0.0377
m_1	0.1627	0.1592
Displacement Direction (d_1)	$\begin{vmatrix} -0.0324 \\ 0.0188 \\ 0.9993 \end{vmatrix}$	$\begin{vmatrix} 0.9979 \\ 0.0195 \\ 0.0615 \end{vmatrix}$
Habit Plane (p_1)	$\begin{vmatrix} 0.9674 \\ 0.0054 \\ 0.2532 \end{vmatrix}$	$\begin{vmatrix} 0.1665 \\ -0.0014 \\ 0.9860 \end{vmatrix}$
Junction Plane (j)	(100)	(001)
Angle p_1,j	14.7°	9.6°
Orientation Relationship $(100)_m{}^\wedge(100)_t$ $(010)_m{}^\wedge(010)_t$ $(001)_m{}^\wedge(001)_t$	0.07° 1.08° 9.04°	8.89° 1.10° 0.09°

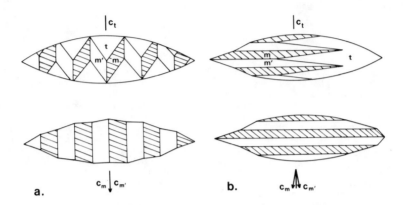

Figure 5   Schematic representation of predicted morphology [16] for m-ZrO$_2$
particles comprising (a) (100)$_m$ twin-related variants, and (b)
(001)$_m$ twin-related variants.   Upper figures illustrate
partially transformed particles.

the habit plane is preserved. It is to be noted that the angle between the
habit plane and the junction plane has been exaggerated in Fig. 5 for clar-
ity; the true angles for the two examples discussed are given in Table 1.

3.3   Implications for Transformation Toughening

The strain around particles such as those represented schematically in
Fig. 5 will be the resultant of the accumulated shape strains for twin-
related monoclinic variants. For both forms of particle observed, it has
been established [16] that the strain component normal to the junction
plane is equal for variants across that plane, while the shear components
of the strain in the junction plane are directly opposed in twin-related
variants. Thus in a transformed particle comprising equal volume fractions
of twin-related variants, the shear components parallel to the junction
plane will exactly cancel. Adjacent twin-related variants are in large
part self-accommodating and the transformation to approximately equal
fractions of twin-related variants minimizes the strain accumulating around
a given particle. The axial strains accumulated parallel to the base vectors of the tetragonal unit cell are shown in Table II, for the four solutions considered in detail by Kelly and Ball [16]. In all cases the strain
parallel to [010]$_t$ (i.e. [010]$_m$) is negligible and the accumulated misfit
strain approximates closely to a plane strain. This model of the transformed product is to be contrasted with that in which the particle is regarded as an internally twinned plate of the monoclinic phase (Fig. 1). In
the latter case, the shape change would be macroscopically homogeneous over
the particle as a whole and the misfit strain would contain large shear and
dilatational components.

A further implication of a transformed particle comprising approximately twin-related variants of the monoclinic phase, is that significant
local strains will be generated at the intersections of the junction plane
with the original interface between tetragonal and cubic phases. For a

Table II   Predicted Axial Misfit Strains for Transformed Particles
            Comprising Equal Fractions of Twin-Related Monoclinic Variants
            [16]

Solution	6A	14A	6B	14B
Axial Strain (%)				
$[100]_t$	-0.51	1.44	0.73	2.65
$[010]_t$	0.00	0.00	0.00	0.00
$[001]_t$	4.12	2.17	2.88	0.97

particle containing transverse $(100)_m$ domain boundaries the displacement is, for a given interface, a maximum along alternate domain boundaries and, for an average domain width of ~ 10nm, amounts to approximately 3 times the interplanar spacing.   In the absence of what have been termed 'closure twins' [26] or significant local plastic deformation, these points of intersection thus likely represent preferred sites for microcrack nucleation [27].   The pattern of microcracking anticipated [17] is very similar to that observed in practice for those particles with transverse $(100)_m$ domain boundaries, Fig. 3.   For those particles with contact plane parallel to $(001)_m$, the local displacements will be associated with points of intersection at the extremities of a plate-like particle and, in particles observed in thin foil specimens, microcracking does appear preferred adjacent to the particle tips [17].

4.   THE STRESS-INDUCED TETRAGONAL-MONOCLINC TRANSFORMATION IN POLYCRYSTALLINE t-ZrO$_2$

In those ceramics containing a microstructure of fine-grained tetragonal phase, the t → m transformation may be stress-activated in, for example, the deformation zone surrounding a hardness indentation or a propagating crack [11,18].   In the immediate vicinity of the stress concentration transformation is essentially complete, but at the perimeter of the deformation zone it is common to observe partially transformed grains comprising one or two variants of the monoclinic phase in retained tetragonal matrix.   Figures 6 and 7 show two examples of such grains in an indented sample of a CeO$_2$-ZrO$_2$ alloy containing 12 mol.% CeO$_2$.   The form of the microstructure in Fig. 6 suggests that the monoclinic phase nucleates preferentially at grain boundaries [11] and those monoclinic variants observed take the form of parallel-sided plates or laths.   Such partially transformed grains provide an excellent opportunity to obtain experimental data for comparison with predictions of the crystallographic theory.

A preliminary crystallographic analysis has been performed for the two grains shown in Figs. 6 and 7 and a third grain of $<100>_t$ orientation, and the results are plotted in stereographic projection in Fig. 8.   Because of the similarity in the a_t and c_t parameters [28] it is difficult to determine unambiguously the c_t axis using electron diffraction and thus lattice correspondence C has been assumed for the purposes of the analysis.   The SAED pattern in Fig. 6(b) has as a result been indexed such that the electron beam is parallel to $[010]_t$ and is in turn parallel to $[010]_m$ for monoclinic variant 1.   The orientation relationship defined in this way is such that $(100)_m//(100)_t$ and $[001]_m//[001]_t$, and this is shown

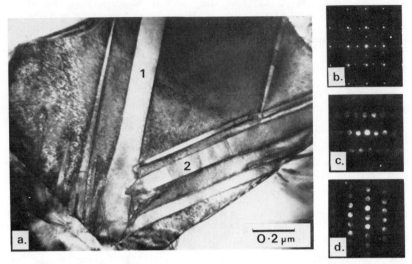

Figure 6. (a) TEM micrograph of partially transformed tetragonal grain in CeO_2-stabilized polycrystalline t-ZrO_2, (b) corresponding $[010]_t$ SAED pattern, and (c), (d) microdiffraction patterns from monoclinic variants 1 and 2 respectively.

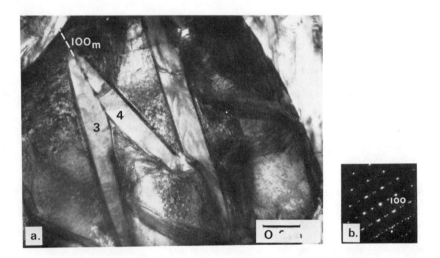

Figure 7 (a) TEM micrograph of partially transformed tetragonal grain in CeO_2-TZP sample, (b) corresponding $[0\bar{1}1]_t$ SAED pattern. Note that junction plane between variants 3 and 4 appears parallel to $(100)_m$.

in Fig. 8. The habit plane of the monoclinic plates has been determined using standard single surface trace analysis and assuming the above variant of the orientation relationship as a standard. For the $<011>_t$ orientation of Fig. 7 the SAED pattern has been indexed such that the electron beam is parallel to $[0\bar{1}1]_t$ and $[0\bar{1}1]_m$ in monoclinic variant 3. In the standard variant of the orientation relationship this is the only orientation in which $<011>$ zone axes of the tetragonal and monoclinic lattices are almost parallel, as the diffraction pattern suggests. The results of trace analysis for the three grains examined are shown in Fig. 8. The trace normals to the habit plane (dashed curves) converge to closely spaced points of intersection near to $(301)_t$. A further interesting result to note is that the two monoclinic variants in Fig. 7 are, at least approximately, twin-related about $(100)_m$ and the trace of the junction plane between them is, within experimental error, parallel to the trace of $(100)_m$.

The experimental orientation relationship shown in Fig. 8 is in excellent agreement with that calculated [16] in solution 6A (Table I) assuming lattice correspondence C. The measured habit planes near to $(301)_t$ are also in good agreement with theoretical predictions for a LIS on $(011)_t$. An average habit plane normal determined from the intersections of trace normals would lie ~ 3° from the habit plane predicted for system 6A (Table I). Together with the observed occurrence of twin-related, and thus partially self-accommodating, variants of the monoclinic phase, these preliminary observations provide encouraging confirmation of the applicability of the crystallographic theory. It is to be stressed, however, that an independent determination of the lattice correpondence and further detailed measurements of the shape strain and orientation relationship are required to provide a proper evaluation of the theoretical predictions.

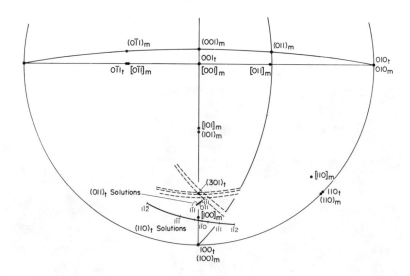

Figure 8 Stereographic projection showing results of measurements of orientation relationship and habit plane for stress-activated t → m transformation in CeO$_2$-TZP. Dashed curves indicate habit plane trace normals for monoclinic variants depicted in Figs. 6 and 7.

It is also to be noted that the predictions of Table I have been determined using lattice parameters for MgO-PSZ and not the CeO$_2$-stabilized materials.

5. FORMATION OF AN ORTHORHOMBIC PHASE IN PSZ

It is now well established that, in addition to the expected tetragonal and monoclinic phases, a metastable orthorhombic phase may form in ZrO$_2$ particles dispersed in thin foils of a range of PSZ samples [19-21] and in discrete single crystal ZrO$_2$ particles prepared by internal oxidation of thin foils of a Nb-Zr alloy [29]. The orthorhombic phase is readily distinguished from both the parent tetragonal phase and the equilibrium monoclinic phase by the appearance of fine-scale, transverse striations within the particles, Fig. 9(a), and the emergence of $\{100\}_o$ and $\{110\}_o$ reflections in $<100>_o$ zone axis diffraction patterns, Fig. 9(b). These reflections are forbidden for the tetragonal lattice [25] and cannot readily be accounted for assuming the monoclinic structure [17].

There is an increasing body of evidence which suggests strongly that this orthorhombic phase is in effect an artefact of specimen preparation and that it only occurs in thin foil specimens used in electron microscopy. This evidence includes observations that:
(i)   there are significant differences in the structure and substructure of the orthorhombic phase formed in ZrO$_2$ particles normal to and parallel to the specimen plane in a thin foil specimen [21],
(ii)   the orthorhombic phase is apparently confined to particles in the thinner areas of a given specimen [21,30] and is not observed in thicker regions of a foil, and
(iii) particles of t-ZrO$_2$ may transform to the orthorhombic phase during ion-milling of thin foil specimens [21]. It appears likely that formation of the orthorhombic phase occurs spontaneously within a particle on thinning of a specimen, perhaps as a result of relaxation of the matrix constraints on the particle.

Although formation of the orthorhombic phase appears of little practical importance, there are aspects of its formation that are interesting fundamentally. The t → o transition appears a displacive reaction and is

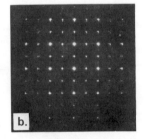

Figure 9   (a) Centred dark field micrograph showing metastable orthorhombic ZrO$_2$ particles in thin foil MgO-PSZ, and (b) corresponding SAED pattern.

Figure 10 TEM micrographs showing
stages in the reverse
transformation of ortho-
rhombic ZrO_2 particle to
tetragonal phase in
MgO-PSZ. Transformation
induced by local electron
beam heating [21].

readily reversible by local heating. The reverse transformation occurs
athermally and the tetragonal phase is only stable when transformation is
complete within the particle [21]. Figure 10 shows an example of a
partially transformed particle, the reverse transformation being induced by
the removal of the condenser aperture of the microscope and controlled
focussing of the electron beam. Figures 10 (a) and (b) compare two
different settings of the second condenser lens and resulting differences
in the location of the o/t interfaces. The reversible transformation
proceeds in either direction by the rapid migration of a planar interface
parallel to $(100)_t$ along the length of the particle in this 'edge-on'
orientation.

The orthorhombic phase formed on thinning of a specimen may in turn
undergo a stress-induced displacive transformation to the equilibrium monoclinic phase, the product of transformation being similar to that of the t
→ m transition occurring in particles dispersed in bulk samples [17].
Figure 11 shows an example of an initially orthorhombic particle which has
partially transformed within the thin foil to monoclinic phase.
Transformation has involved the growth of just two self-accommodating
variants of the monoclinic phase along the centre of the particle, the
variants being approximately twin-related about $(001)_m$ [17]. The transformation was observed to initiate at one end of the particle and propagate
rapidly along the length; the regions along the edges of the particle
remain orthorhombic [21]. The traces of the interfaces between the
monoclinic laths and the orthorhombic phase deviate ~ 1-2° from the traces
of the $(001)_m$ plane (or $(001)_o$). The orientation relationship between
orthorhombic and monoclinic lattices is such that $(010)_o//(001)_m$ and
$[001]_o//[010]_m$ [21]. A similar, displacive o → m transformation has been
observed in thin single crystal ZrO_2 particles prepared by the internal
oxidation method [29]. In this case the orientation relationship is
reported as $(100)_o//(100)_m$ and $[001]_o//[001]_m$, and the habit plane is

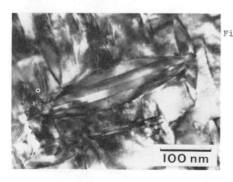

Figure 11 TEM micrograph showing
large orthorhombic ZrO_2
particle partially trans-
formed to monoclinic
phase. Edges of the
particle remain
orthorhombic.

described as being approximately $(80\bar{1})_o$, deviating 13-14° from the $(100)_o$ plane.

Although occurring in a thin foil specimen and thus perhaps atypical of bulk transformation, the structure of the partially transformed particle of Fig. 11 does support strongly the view that a transformed monoclinic particle is to be regarded as comprising self-accommodating, twin-related variants of the monoclinic phase.

References

1. A.H. Heuer and L.W. Hobbs (Editors), Advances in Ceramics, Vol. 3, Science and Technology of Zirconia. Am. Ceram. Soc., Columbus, OH, 1981.
2. N. Claussen, M. Rühle and A.H. Heuer (Editors). Advances in Ceramics, Vol. 12, Science and Technology of Zirconia II. Am. Ceram. Soc., Columbus, OH, 1984.
3. A.G. Evans, ibid., pp.193-212.
4. R.H.J. Hannink and M.V. Swain, J. Mat. Sci. Letters, 16 1980 (1981).
5. J.E. Bailey, Proc. Roy. Soc. London, Ser. A, A279 395 (1964).
6. G.K. Bansal and A.H. Heuer, Acta Metall., 20 1281 (1972); 22 409 (1974).
7. E.C. Subbarao, H.S. Maiti and K.K. Srivastava, Phys. Status Solidi A, 21 9 (1974).
8. W.M. Kriven, W.L. Fraser and S.W. Kennedy, in ref. [1], pp.82-97.
9. M.A. Choudry and A.G. Crocker, in ref. [2], pp.46-53.
10. S.T. Buljan, H.A. McKinstry and V.S. Stubican, J. Am. Ceram. Soc., 59 351 (1976).
11. M. Rühle and A.H. Heuer, in ref. [2], pp.14-32.
12. G.K. Bansal and A.H. Heuer, J. Am. Ceram. Soc., 58 235 (1975).
13. D.L. Porter and A.H. Heuer, J. Am. Ceram. Soc., 62 298 (1979).
14. W.M. Kriven, in ref [1], pp.168-83.
15. W.M. Kriven, in ref [2], pp.64-77.
16. P.M. Kelly and C.J. Ball, J. Am. Ceram. Soc., 69 259 (1986).
17. B.C. Muddle and R.H.J. Hannink, J. Am. Ceram. Soc., 69 547 (1986).
18. R.H.J. Hannink, B.C. Muddle and M.V. Swain, Austceram 86, Proc. 12th Aust. Ceramic Conf., (Australian Ceramic Soc., 1986), pp.145-152.
19. L.K. Lenz and A.H. Heuer, J. Am. Ceram. Soc., 65 C192 (1982).
20. A.H. Heuer, L.H. Schoenlein and S. Farmer, in Science of Ceramics, Vol. 12, edited by P. Vincenzini (Ceramurgica s.r.l. Faenza, Italy), pp.257-66.
21. B.C. Muddle and R.H.J. Hannink, Proceedings Zirconia '86, Tokyo, Japan, submitted for publication.
22. L.H. Schoenlein and A.H. Heuer, in Fracture Mechanics of Ceramics, Vol. 6, edited by R.C. Bradt, A.G. Evans, D.P.H. Hasselman and F.F. Lange (Plenum, New York, 1983), pp.309-25.
23. J.S. Bowles and J.K. Mackenzie, Acta Metall., 2 129 (1954); 2 138 (1954).
24. M.S. Wechsler, D.S. Lieberman and T.A. Read, Trans. Am. Inst. Min. Eng., 197 1503 (1953).
25. R.H.J. Hannink, J. Mat. Sci., 13 2487 (1978).
26. E. Bischoff and M. Rühle, J. Am. Ceram. Soc., 66 123 (1983).
27. Y. Fu, A.G. Evans and W.M. Kriven, J. Am. Ceram. Soc., 67 626 (1984).
28. K. Tsukuma and M. Shimada, J. Mat. Sci., 20 1178 (1985).
29. Y.-H. Chiao and I.-Wei Chen, to be published in Proc. JIMIS-4, Grain Boundary Structure and Related Phenomena, Suppl. to Trans. Jap. Inst. Metals, (1985).
30. R.M. Dickerson, M.V. Swain and A.H. Heuer, J. Am. Ceram. Soc., in press (1986).

A STUDY OF RHOMBOHEDRAL PHASE
IN Y_2O_3-PARTIALLY STABILIZED ZIRCONIA

YUKISHIGE KITANO[*], Y. MORI[*], A. ISHITANI[*] AND T. MASAKI[**]
[*] Toray Research Center, Inc., Otsu, Shiga, 520, Japan
[**] Technical Development Department, Toray Industries, Inc.,
Otsu, Shiga, 520, Japan

ABSTRACT

Tetragonal to rhombohedral phase transformation was studied
by X-ray diffraction technique on the ground surfaces of tetragonal zirconia polycrystals (Y-TZP) and partially stabilized
zirconia (Y-PSZ) with 2.0 to 5.0 mol% Y_2O_3 contents prepared by
hot isostatic pressing. The rhombohedral phase increased
with increase of Y_2O_3 content from 2.0 to 5.0 mol%, and also
with the increase of HIPing temperature from 1400 to 1600°C.
The stability of the phase was also studied with regard to the
surface finish and annealing. The subsequent heat treatment of
the samples was found to promote the reverse rhombohedral to
tetragonal transformation.

INTRODUCTION

It is well-known that zirconia has three allotropes: monoclinic, tetragonal and cubic phases [1]. Transformation from
tetragonal to monoclinic phase has been extensively studied in
conjunction with enhancement of toughness and strength in tetragonal zirconia polycrystals [2-8]. Recently, the existence of
an additional new phase, rhombohedral, was reported [9-11]. The
rhombohedral phase was formed from the cubic phase through the
transformation under the stress not only on the abraded surface
but also on the ion-implanted surface of both partially stabilized and fully stabilized zirconias [9,10]. It was also indicated that the rhombohedral phase appeared in the arc-melted
bulk of ZrO_2-Y_2O_3 alloys with the composition of 3 to 4 mol%
Y_2O_3 without any mechanical treatment [11].
The present work reports the existence of rhombohedral
phase in Y_2O_3 containing partially stabilized zirconia and
tetragonal zirconia polycrystals with 2.0 to 5.0 mol% Y_2O_3
contents prepared by hot isostatic pressing (HIP).

EXPERIMENTAL

Polycrystalline ZrO_2's stabilized with 2.0 to 5.0 mol% Y_2O_3
containing 0.5 mol% Al_2O_3 were prepared. A mixed solution of
hydrochlorides was prepared as the starting material.
The zirconia powder was prepared by thermal decomposition of the
solution followed by calcining and grinding by the method previously reported [12].
The test specimens were prepared by hot isostatic pressing.
The powder was first isostatically pressed at 200 MPa and
ambient temperature and then heated to 900°C at a rate of 50°C/h,
from 900°C to sintering temperature at 30°C/h and finally held
at 1350 to 1450°C for 2h. Then an uncapsulated sample was
prepared under argon containing oxygen of about 4 mol% or under

argon with increasing pressure up to 200 MPa at room temperature: the temperature was then raised to between 1350 and 1500°C at a rate of 700°C/h and held for 1.5h. The sintered specimens thus obtained were then ground at first with 400-grit diamond wheel and then with 10μm diamond paste to optical finish. To investigate the effect of surface finish on the phase composition, two sets of specimens HIPed at 1450°C under argon containing oxygen of 4 mol% were selected for X-ray analysis; one was Y-TZP containing 2.5 mol% Y_2O_3 and the other Y-PSZ containing 5.0 mol% Y_2O_3. The surfaces of two sets of specimens received further polishing with 5μm and 3μm diamond paste. The fracture surfaces produced in bending tests were also subjected to X-ray analysis for phase composition determination. Thermal annealings of the ground specimens of the same sets were conducted at 600, 800 and 1000 °C in air for 24h.

The X-ray diffraction measurements were performed by the step-scanning method with graphite monochromated Cu Kα radiation. The X-ray diffraction lines of the rhombohedral phase of the specimens tested appeared on the low angle side of every tetragonal line. The lattice parameters of the rhombohedral phase were determined by the least squares fitting of six peaks obtained over a range of 2θ=30-73°after Kα$_1$ and Kα$_2$ peak separation and multipeak resolution. The proportions of the cubic, tetragonal and monoclinic phases were estimated from the relative areas under (11$\bar{1}$)m, (111)m and (111)t,c, and (004)t, (400)t and (400)c X-ray diffraction profiles using the method of Garvie and Nicholson [13]. The fraction of the rhombohedral phase was not quantitatively evaluated, since its X-ray diffraction lines overlapped those of the tetragonal and cubic phases.

RESULTS AND DISCUSSION

Figure 1 shows the X-ray diffraction profiles for the ground surfaces of Y-TZP and Y-PSZ specimens containing 2.0 to 4.0 mol% Y_2O_3 prepared by HIPing temperature of 1400 and 1600°C under argon, respectively. Inspection of the peaks indicates that the asymmetry of the profiles of the tetragonal/cubic (111) lines becomes larger with increase of Y_2O_3 content and a new component can be clearly recognized in 4.0 mol% Y-PSZ HIPed at 1600 °C. This new diffraction component which appears on the low angle side of every tetragonal line could be assigned to be due to the rhombohedral phase.

Lattice parameters of the new diffraction component were determined from the diffraction lines of the sample with 4.0 mol% Y_2O_3 HIPed at 1600°C, which contains the largest amount of the new phase in the series of tested specimens. The least squares fitting procedure after Kα$_1$ and Kα$_2$ separation and the multipeak resolution by a method of a linear combination of Gaussian and Cauchy functions [16] shown in Fig. 2 has yielded the following rhombohedral parameteres:
a=0.3664 nm, α=59.9°(Z=1); a'=0.5181 nm, α'=89.9°(Z=4) and a$_h$=0.3661 nm, c$_h$=0.898 nm (Z=3) by the hexagonal notation. These values are in good agreement with those reported previously for abraded and ion-implanted surfaces, and also for the arc-melted bulk of Y_2O_3-containing zirconia [9-11], and moreover can be distinguished from the orthorhombic phases (JCPDS card No. 34-1084 and 33-1483). Thus the crystal structure of the rhombohedral zirconia in the present study is considered to be

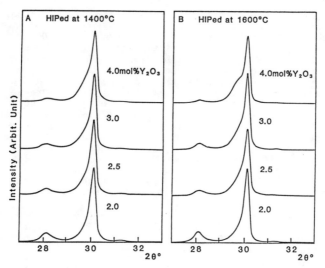

Figure 1.   X-ray diffraction profiles for the ground surfaces of Y-TZP abd Y-PSZ with 2.0 to 4.0 mol% Y_2O_3 contents HIPed at 1400° and 1600°C, respectively.

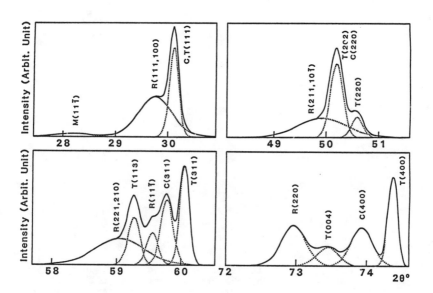

Figure 2.   Multipeak resolution of X-ray diffraction profiles after $K\alpha_1$ and $K\alpha_2$ separation.

exactly the same as one formed from the cubic phase through transformation under the stress.

Influence of the HIPing temperature between 1350 and 1600°C on appearance of the new rhombohedral phase for the ground surfaces of 2.5 mol% Y-TZP is shown in Fig. 3. The rhombohedral phase is found to increase with increase in HIPing temperature. Volume fractions of cubic, tetragonal and monoclinic phases of the specimens are summarized in Table I. With increase in Y_2O_3 content and in HIPing temperature, a continuous increase of cubic phase with accompanying decrease of tetragonal phase is observed. These observations can be explained using the phase diagram, where the cubic phase becomes more stable than the tetragonal one at higher sintering temperature and with higher content of Y_2O_3 [14,15]. Decrease in the monoclinic phase may be attributed to decrease of the tetragonal phase in the specimens used. It could be inferred from these results that the rhombohedral phase was formed on the ground surfaces of Y-TZP and Y-PSZ containing higher amount of tetragonal phase. The increase in the rhombohedral phase appears to be consistent with the increase in cubic phase.

Effect of surface finish and thermal annealing on the phase contents was investigated. The results are shown in Fig. 4 and Table II. The as-sintered specimens are free from rhombohedral phase, but it is generated by grinding; the process seems to induce high stress levels on the surface. When the ground surfaces are subjected to successive finer polishing, the rhombohedral phase disappears. At the same time the monoclinic phase also disappears from the surface of the polished specimens. This indicates that the disappearance of rhombohedral phase is the result of removal of the outer strained thin surface layer and that only the cubic and tetragonal phase are existing in the bulk of the specimens. As the content of the cubic phase changes little by surface finish, it can be suggested that the rhombohedral phase appeared on the

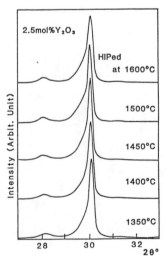

Figure 3. X-ray diffraction profiles for the ground surfaces of 2.5 mol% Y_2O_3-TZP HIPed at 1350, 1400, 1500 and 1600°C, respectively.

Table I. Effect of sintering conditions on the phase
contents of ground surfaces of Y-TZP and Y-PSZ

Y_2O_3 content (mol%)	Sintering temperature PS**/HIP (°C)	Phase contents(%)*		
		C	T	M
2.0	1450/1400	–	87	13
2.5	1450/1400	5	88	7
3.0	1450/1400	8	86	6
4.0	1450/1400	20	77	3
2.0	1450/1600	–	86	14
2.5	1400/1600	10	83	7
3.0	1450/1600	13	83	4
4.0	1400/1600	26	71	3
2.5	1350/1350	–	95	5
2.5	1350/1400	–	94	6
2.5	1350/1450	–	94	6
2.5	1450/1500	6	87	7
2.5	1400/1600	10	83	7

* C:cubic, T:tetragonal, M:monoclinic
**PS:pressureless sintering

ground surfaces is formed from tetragonal phase through transformation under mechanical stress induced on the outer surfaces.
    Further observation was conducted on the fracture speci-
mens. Increase in monoclinic phase and decrease in tetragonal
phase are observed as a result of tetragonal to monoclinic
transformation in the specimens, but the rhombohedral phase is
not detected.
    The ground specimens were then subjected to an additional
thermal annealing in air for 24h. After annealing at 600 and
800 °C, the rhombohedral phase decreased. As shown in Fig. 4
and Table II, decrease in rhombohedral content and increase in
monoclinic phase with small change in cubic phase indicate that
the rhombohedral to monoclinic transformation could be operative
by annealing at this range of temperature (600 - 800°C). The
amount of tetragonal phase content is really related to amount
of cubic plus monoclinic phase (tetragonal phase will decrease
by transformation to monoclinic phase). At the same time,
decrease in rhombohedral phase with annealing appears to correspond with the initial increase in monoclinic phase, especially
in 2.5 mol% Y_2O_3 material. The rhombohedral to monoclinic transformation appears to occur during annealing. With increase in
annealing temterature, the monoclinic to tetragonal transformation should occur.   When annealed at low temperature, the
rhombohedral to monoclinic transformation can occure; thus
relative amount of monoclinic phase increases.   At higher
annealing temperature the monoclinic to tetragonal transformation occurs. With increase in Y_2O_3 content (5 mol%), however,
the monoclinic to tetragonal and rhombohedral to tetragonal
transformation during annealing will occur more easily than in
the 2.5 mol% material.   In fact the monoclinic to tetragonal

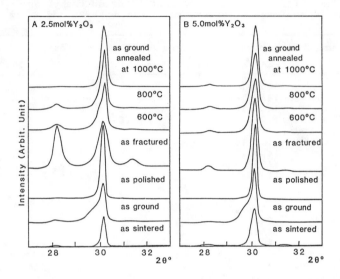

Figure 4. Effect of surface finish and thermal annealing
on the 2.5 mol% Y-TZP and 5.0 mol% Y-PSZ crystal structure.

Table II. Effect of surface finish and thermal annealing
on the phase contents

| | 2.5 mol% Y_2O_3-TZP | | | 5.0 mol% Y_2O_3-PSZ | | |
| | Phase contents (%)* | | | Phase contents (%)* | | |
	C	T	M	C	T	M
as-sintered	11	86	3	50	47	3
as-ground	12	84	4	52	47	1
as-polished	12	88	–	53	47	–
as-fractured	10	43	47	55	37	8
as-ground annealed						
at 600°C	10	80	10	41	54	5
800°C	11	79	10	42	53	5
1000°C	10	89	1	43	55	2

* C:cubic, T:tetragonal, M:monoclinic

transformation could occur below 600°C. Thus the analysis of the X-ray data would indicate that annealing at 600 °C and above results in a decrease in cubic phase content relative to the increase in the tetragonal phase (Table II, 5 mol% Y_2O_3). Table II also indicates that the rhombohedral to tetragonal transformation will occur. After annealing at 1000 °C, the rhombohedral phase disappears. Since cubic phase content does not increase at higher temperature and tetragonal phase content increases during annaling compared with as ground materials, rhombohedral phase will transform reversibly to the tetragonal phase by the thermal annealing at 1000 °C.

CONCLUSIONS

Tetragonal to rhombohedral stress-induced phase transformation was found by X-ray diffraction technique on the ground surfaces of tetragonal zirconia polycrystals and partially stabilized zirconias with 2.0 to 5.0 mol% Y_2O_3 contents prepared by hot isostatic pressing. The rhombohedral phase increased with Y_2O_3 contents and also with HIPing temperature. The stability of the rhombohedral phase was also studied with regard to the surface finish and the thermal annealing. The rhombohedral phase is generated by rough grinding on surface, and decreases with finer polishing and also elevation of temperature.

ACKOWLEDGEMENTS

The authors would like to express their thanks to Prof. I-Wei Chen of Michigan University and Mr. K. Nakajima of Toray Industries, Inc. for their helpful discussions. Thanks are also due to Mr. T. Hori of Rigaku Denki Corporation for the technical assistance of X-ray multipeak resolution and Mr. Y. Ueda of Toray Industries, Inc. for the sample preparation.

REFERENCES

1. R.C. Garvie, R.H. Hannink and R.T. Pascoe, Nature 258, 703 (1975).

2. T.K. Gupta, J.H. Bechtold, R.C. Kuznicki, L.H. Cadoff and B.R. Rossing, J. Mater. Sci. 12, 2421 (1977).

3. T. Masaki and K. Kobayashi, in Proceedings of the Japanese Ceramics Society Meeting (1981), pp. 2-3.

4. K. Tsukuma, Y. Kubota and T. Tsukidate, in Advances in Ceramics, Vol. 12, Science and Technology of Zirconia II, edited by N. Claussen, M. Rühle and A.H. Heuer (American Ceramic Soc., Columbus, Ohio, 1984) pp. 382-390.

5. R. M. McMeeking and A.G. Evans, J. Am. Ceram. Soc. 65, 242 (1982).

6. F.F. Lange, J. Mater. Sci. 17, 225-263 (1982).

7. A.G. Evans and A.H. Heuer, J. Am. Ceram. Soc. 63, 242 (1982).

24

8. V.K. Pujari and I. Jawed, J. Am. Ceram. Soc. <u>68</u>, C242 (1985).

9. H. Hasegawa, J. Mater. Sci. Lett. <u>2</u>, 91 (1983).

10. H. Hasegawa, T. Hioki and O. Kamigaito, J. Mater. Sci. Lett. <u>4</u>, 1092 (1985).

11. T. Sakuma, Y. Yoshizawa and H. Suto, J. Mater. Sci. Lett. <u>4</u>, 29 (1985).

12. T. Masaki, J. Am. Ceram. Soc. <u>69</u>, 519(1986); <u>69</u>, 638 (1986).

13. R.C. Garvie and P.S. Nicholson, J. Am. Ceram. Soc. <u>55</u>, 303 (1972).

14. H.G. Scott, J. Mater. Sci. <u>10</u>, 1527 (1975).

15. A.H. Heuer and M.Rühle, in <u>Advances in Ceramics, vol.12, Science and Technology of Zirconia II</u>, edited by N. Claussen, M. Ruhle and A.H. Heuer (American Ceram. Soc., Columbus, Ohio, 1984), pp. 1-13.

16. A.M. Hindeleh and D.J. Johnson, J. Phys. D. Appl. Phys. <u>4</u>, 259 (1971); Polymer <u>13</u>, 423 (1972).

TEXTURE ON GROUND, FRACTURED, AND AGED Y-TZP SURFACES

FRANZ REIDINGER AND PHILIP J. WHALEN
Allied-Signal, Inc., P. O. Box 1021-R, Morristown, NJ 07960

ABSTRACT

The phase composition of Y-TZP surfaces has been shown to vary greatly depending on the thermo-mechanical history of the surface. The orientation of these different phases in the surface region is not always random. There is speculation that the alignment of the tetragonal phase before fracturing may play a part in increasing the toughness of these materials. This article deals with an X-ray diffraction analysis of various Y-TZP surfaces with special emphasis on the texture of the different phases. Surfaces which have been ground (and polished), fractured, and aged (200°C) have been examined. In all cases, the monoclinic component that was formed was strongly oriented. The tetragonal phase may or may not be oriented depending on surface treatment. Annealing above the monoclinic-tetragonal transition temperature had little effect on the tetragonal orientation in most cases. Samples fractured at 1000°C have no unusual orientation on the fracture faces.

1. INTRODUCTION

The stress induced tetragonal-monoclinic phase transformation, in ceramics containing zirconia as either the major or minor component, is an established principal toughening mechanism. This transformation has been studied extensively [1,2] and is the basis for a growing group of materials called transformation toughened ceramics. Models exist which correlate mechanical properties (toughness, strength) with the degree of transformation. Diffraction patterns of individual crystals around cracks (TEM) and fracture surfaces provide the evidence that the tetragonal to monoclinic transformation occurs on fracturing. There is some debate, however, as to whether the transformation alone can account for the high strength and toughness found in purely tetragonal (polycrystalline or single crystal) zirconia materials. Virkar and Matsumoto [3] have compiled a set of examples from the literature where the toughness of tetragonal zirconia could not be attributed to the tetragonal to monoclinic transformation. It is their contention that a large part of the toughness of these materials is due to a ferroelastic domain switching mechanism. The enhancement of certain peaks in XRD patterns of ground surfaces of TZP (as well as established partially ferroelastic materials like $BaTiO_3$) was given as evidence that reorientation of ferroelastic domains in tetragonal zirconia occurs. Except for the above study, no data are available on the orientation of phases on actual fracture faces or other TZP surfaces. In this paper, we present a detailed X-ray diffraction analysis of yttria-TZP surfaces, including fracture faces.

2. ANALYSIS PROCEDURE

Sintered, dense (>98%) Y-TZP * samples were used for this study. Toyo Soda material was chosen due to its low silica content and resistance to yttria migration on high temperature annealing [4]. Ground surfaces were obtained by machining with a 180 M diamond wheel. Polished surfaces were finished with diamond paste (3 μ in. finish). Low temperature aging studies were done at 200°C in air. High temperature anneals were also done in

* TZ-2.5Y - 2.5 mole % Y_2O_3/ZrO_2 - Toyo Soda, Inc., Atlanta, Georgia

Mat. Res. Soc. Symp. Proc. Vol. 78. ⸱1987 Materials Research Society

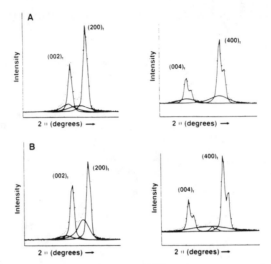

Fig. 1. Decomposed X-ray diffraction patterns of
Y-TZP fired at (A) 1325°C and (B) 1550°C.

ambient atmospheres. The fracture faces were cut from 4 pt. bend samples
(6mm x 6.5mm x 75mm) which were indented with a Vickers diamond on the tensile face. A Philips diffractometer ADP 3600 with a graphite exit beam
monochromator and Cu K_α radiation at 45 KV and 40 mA was used for collecting
the diffraction data. The diffraction patterns were analyzed with the peak
fitting program QPROF which uses modified Lorentzians to describe the peak
shape.

3. RESULTS

The crystalline phases of these nominally 4.45 wt% Y_2O_3/ZrO_2 materials
depend strongly on the firing temperature and any subsequent anneals. Fig.
1 shows a typical peak profile analysis on two polished samples fired to
different temperatures. The patterns from the 200 and 400 regions, contain
both sharp and broad components. For samples fired at low temperature
(1325°C) this broad component was mainly due to strain. The broad component
between the 200 and 002 peaks, increased significantly on samples fired at
higher temperatures (1550°C). Also, a small sharp peak developed between
the 400 and 004 reflections. This additional scattering intensity, between
the 200/002 and 400/004 peaks was likely due to the t'-ZrO_2 phase which has
been identified in quenched 4 wt% YZr [5] and 6 to 12 wt% YZr [6,7].

3.1 Tetragonal texture on treated surfaces

Two distinct types of orientation were observed on the Y-TZP surfaces
due to effects of grinding and annealing, Fig. 2. The first type is similar
to the texture developed in rolled metals, where the individual crystals
orient (or grow) in a preferred direction. In the case of a ground and
annealed Y-TZP surface, the XRD reflections in the 200 region were strongly
enhanced, Fig 2a. Table 1 contains the analysis of XRD patterns of polished
and ground surfaces after different treatments. The enhancement of the 200
region required annealing of a ground surface at elevated temperatures.

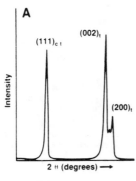

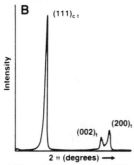

Fig. 2 XRD patterns for Y-TZP surfaces which were (A) ground and (B) polished before annealing at 1550°C for 20 hours.

The second type of preferred orientation, namely an increase in the I_{002}/I_{200} ratio, which has been previously reported [3], was induced by grinding alone. This effect is illustrated by examining the I_{002}/I_{200} ratios in the XRD patterns of Fig. 2, which are tabulated in Table I. The "standard" and polished samples have I_{002}/I_{200} ratios close to 0.6, whereas the ground surface, both before and after annealing, showed a large enhancement of the (002) reflection and a I_{002}/I_{200} ratio of about 2. The observed ratios on the lower firing temperature surfaces were close to the ideal case. The presence of the broad component affected the ratios in the higher temperature samples. In general the low angle component of related reflections was enhanced during grinding. For higher order reflections, i.e. 400, the increase was reduced, because only a thin layer was affected by grinding. This effect was not a transitory strain state as the enhancement remained even after 20 hours at 1550°C.

3.2 Monoclinic texture on Y-TZP surfaces

The tetragonal to monoclinic transformation in Y-TZP can be induced by grinding [8], aging at low temperatures [9], and etching (boiling H_2SO_4). An XRD analysis of surfaces so treated indicated that a strongly oriented monoclinic phase developed. Similar to the tetragonal case discussed above, the log angle component of related reflections was enhanced, i.e., the $I_{11\bar{1}}/I_{111}$ in Fig. 3a&b. In some cases where the transformation was not complete, the high angle component may not be detectable, like in the (011)-(110) and (002)-(200) reflections, Fig. 4. The tetragonal texture (I_{002}/I_{200} ~1) appears to be affected when the T-M transformation was sufficiently advanced. In the 400 region, which is relatively free of interference from monoclinic peaks, the tetragonal 004/400 ratio remained about 0.4.

3.3 Texture on fracture faces

Many of the characteristics of the phases found on treated surfaces were also found on the fracture faces of Y-TZP bars, as shown in Fig. 5 and Table II. The tetragonal texture varied depending on the amount of monoclinic formed. As the monoclinic increased, the (002) reflection of the tetragonal decreased. The result is an apparent enhancement of the (200) reflection, which was opposite to the orientation found on ground surfaces. A large broad peak contributes to the intensity of the 200 region, and a sharp peak (probably t'-ZrO_2) was present in the 1550°C pattern.

28

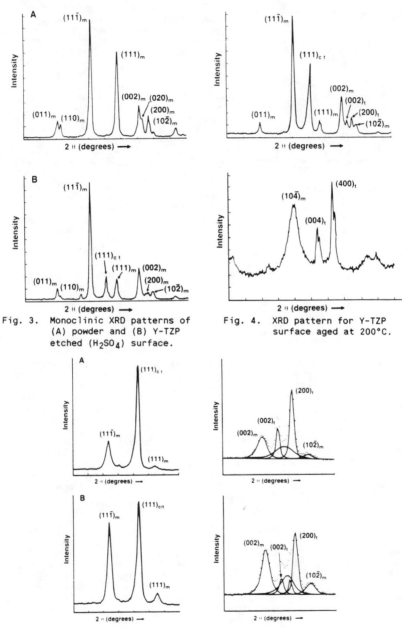

Fig. 3. Monoclinic XRD patterns of (A) powder and (B) Y-TZP etched (H₂SO₄) surface.

Fig. 4. XRD pattern for Y-TZP surface aged at 200°C.

Fig. 5. XRD patterns for Y-TZP fracture faces of samples fired at (A) 1375°C and (B) 1550°C.

Table I. Analysis of Intensities of the (200) and (400) Regions
of Tetragonal XRD Patterns

Firing Temp, °C	Treatment	I_{002}	I_{200}	200[a] Region	I_{004}	I_{400}	400[a] Region
	Standard[c]	40	60	20	29	71	7
1325	Polished	25/12	45/-	25	19/11	49/21	14
	Ground	52/17	23/-	25	47/(15)	22/16	12
	Fractured @ 1000°C	32/6	48/14	24	22/12	45/21	15
1550	Polished	35	37/28	24	21/-	49/-	15
	Ground	38/47	15	24	47/-	13/27	10
	Polished and Annealed 20 h @ 1550°C	35	45/20	28	24/-	56/-	15
	Ground and Annealed 20 h @ 1550°C	62/-	17/-	120[b]	46/-	23/-	36
	Fractured @ 1000°C	37	38/25	25	26	53/(21)	15

N_1/N_2 Peak consists of two compoments: $FWHH(N_1) < FWHH(N_2)$.

$N_1/(N_2)$ Slight overlap between broad 00L and H00 peaks.

$N_1/-$ Severe overlap of broad component; cannot be assigned to 00L or H00. Sum of intensities is less than 100.

N   Only one component.

[a] Integrated intensity of (200) and (400) region. (111) region normalized to 100.

[b] The integrated intensity of the (111) was reduced by 35%. For normalization, the (111) intensity of the polished side was used.

[c] Computer generated values using data from D. Michel, L. Mazerolles, M. Perex Y Jorba, J. Mat. Sci., 18 (1983), 2618-2628. The calculated intensities depend on the atomic parameters.

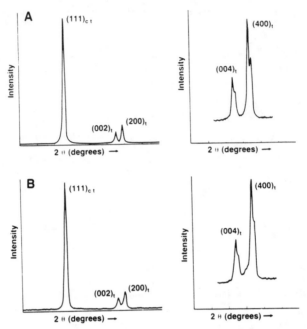

Fig. 6. XRD patterns of fracture faces of samples broken at 1000°C: (A) fired at 1325°C; (B) fired at 1550°C.

Table II. Analyses of Selected XRD Intensities from Y-TZP Fractured Faces

Sample	Monoclinic		Tetragonal*	
	% Monoclinic	$I_{11\bar{1}}/I_{111}$	I_{002}/I_{200}	I_{004}/I_{400}
Standard	100	1.5	0.66	0.40
Fired @ 1325°C				
R.T. Fracture	30	7.4	0.41	0.33
1000°C Fracture	0	---	0.67	0.49
Fired @ 1550°C				
R.T. Fracture	47	6.5	0.44	0.28
1000°C Fracture	0	---	0.97	0.49

* Only narrow peaks were used; see Table I.

In addition, the broadening of the tetragonal peaks (especially the low angle components) was significantly less on fracture faces than that produced during grinding. Unlike a ground surface, the fracture face after a total anneal, in general, regained its original phase composition and texture. At intermediate stages of annealing, however, there are some interesting changes in the amount and texture of both phases, which are under further investigation.

The dependence of the tetragonal orientation on the amount of monoclinic present on fracture faces was studied further with high temperature fracture samples. Bars were broken at 1000°C which is above the transformation temperature and the resultant fracture faces were analyzed, Fig. 6. If a ferroelastic domain switching mechanism was operating during fracture of this material, one would expect to see an intensified effect at this higher temperature because it is closer to the ferroelastic transition point (cubic to tetragonal phase boundary). Also, there should be no complicating effects from a monoclinic phase. The results show that no monoclinic was present and the orientation of the tetragonal phase was identical to polished sections of these samples.

4. DISCUSSION

The texture analysis on Y-TZP surfaces was complicated by the presence of strain broadening and peaks due to second phases in the crucial 200 region. The t'-ZrO_2 phase played a significant role in samples fired at high temperatures (1550°C). The presence of a rhombohedral phase was also considered. Two rhombohedral phases have been proposed, by Hasegawa [10] and Kitano et al. [11], which could have contributed to the enhancement of the 002_t peak. However, no peak from these two phases falls between the 002_t and 200_t reflections.

An indication of the stress state of the fracture surface can be obtained from an analysis of the monoclinic texture. Matsui, et al. [12] have investigated the effect of tensile and compressive stresses on the texture of the monoclinic phase. They report that the average monoclinic $I_{11\bar{1}}/I_{111}$ ratios were 3 to 1 and 10 to 1 for surfaces under tension and compression, respectively. We found that the $I_{11\bar{1}}/I_{111}$ ratio from the fracture faces was about 7 to 1 (Table II) which agrees with the ratios obtained from our aged (Fig. 4a), etched (Fig. 3b), and the as-prepared sample of Matsui et al. The monoclinic texture on fracture faces appears to be due more to accommodation of the anisotropic strain of the monoclinic transformation than to the stresses present on the fracture face.

The texture of the tetragonal phase was more susceptible to surface treatments than the monoclinic phase. Virkar and Matsumoto [3] proposed that the texture of the tetragonal phase on the fracture face should be similar to that found on grinding (i.e., I_{00L} increases). Our data show that the reverse enhancement (i.e., a relative increase of the I_{H00}), was present on the fracture face. While this observation contradicts the prediction of Virkar and Matsumoto, it does not rule out the possibility that a ferroelastic domain switch had occurred. However, the invariance of the I_{002}/I_{200} tetragonal ratio on the 1000°C fracture faces, from the standard (Table II), presents a strong argument against a ferroelastic switching mechanism. The tetragonal texture that was observed on R.T. fracture faces, can then be rationalized by assuming that the monoclinic phase forms preferentially at the expense of grains oriented in the 00L direction.

5. SUMMARY

The surface phases in Y-TZP were analyzed by XRD with special emphasis on the texture developed during thermo-mechanical treatment. These data were then used as a basis to analyze the phases developed on fracture faces. The following observations were made:

a.   Two types of preferred orientation were observed on Y-TZP surfaces; grinding enhances the lower angle component of related tetragonal reflections and subsequent annealing at high temperatures enhances the entire 200 region.

b.   The monoclinic phase formed on Y-TZP surfaces such that the lower angle component of related reflections was enhanced.

c.   Tetragonal phase orientation on fracture faces was opposite of that found on ground surfaces and was dependent on monoclinic content.

d.   No orientation of the tetragonal phase was observed on fracture faces from samples broken at 1000°C. Therefore, it is unlikely that an irreversible tetragonal ferroelastic domain switching mechanism, as proposed by Virkar and Matsumoto [3], is operating during fracture of 4.5 wt % Y-TZP.

ACKNOWLEDGMENTS

We would like to thank J. Van Ackeren for his assistance in fracturing samples at high temperature, S. T. Correale for XRF analysis, H. Minor for his contribution to the XRD studies, and Diane Rusconi and Nina Kim Akgunduz for sample preparation. Finally, we would like to thank Marcy Daboul for accommodating our unreasonable requests during preparation of this manuscript.

REFERENCES

1.   Advances in Ceramics, Vol. 3: Science and Technology of Zirconia, Edited by A. H. Heuer and L. W. Hobbs (The American Ceramic Society, Columbus, Ohio, 1981).

2.   Advances in Ceramics, Vol. 12: Science and Technology of Zirconia II, Edited by N. Claussen, M. Ruhle and A. H. Heuer (The American Ceramic Society, Columbus Ohio, 1984).

3.   A. Virkar and R. Matsumoto, J. Am. Ceram. Soc., 69 [10], C-224-C-226 (1986).

4.   P. J. Whalen, S. T. Correale and F. Reidinger, presented at the 1986 American Ceramic Society Pacific Coast Meeting, Seattle, WA, 1986 (unpublished).

5.   R. Chaim, D. G. Brandon and A. H. Heuer, Acta. Metall., Vol. 34, No. 10, 1936-1939 (1986).

6.   A. Paterson and R. Stevens, J. Mater. Res. 1 (2), Mar/Apr, 295-299, (1986).

7.   V. Lanteri, R. Chaim and A. H. Heuer, J. Am. Ceram. Soc., 69 [10], C-258-C-261, (1986).

8. J. S. Reed and A. Lejus, Mater. Res. Bull., 12 [10], 949-954, (1977).

9. M. Watanabe, S. Iio and I. Fukura, Advances in Ceramics, Vol. 12,
    Edited by N. Claussen, M. Ruhle and A. H. Heuer (American Ceramic
    Society, Columbus, Ohio, 1983) pp. 391-398.

10. H. Hasegawa, J. Mater. Sci. Lett., 2 (1983) 91.

11. Y. Kitano, Y. Mori, A. Ishitani and T. Masaki, "A Study of Rhombohedral
    Phase in Y_2O_3 Partially Stabilized Zirconia", this volume.

12. M. Matsui, T. Soma, and I. Oda, J. Am. Ceram. Soc., 69 [3], 198-202
    (1986).

CONSTITUTIVE LAWS FOR CERAMICS EXHIBITING
STRESS-INDUCED MARTENSITIC TRANSFORMATION

JOHN C. LAMBROPOULOS
Department of Mechanical Engineering, University of Rochester, Rochester,
New York 14627, U.S.A.

ABSTRACT

The theory of internal variables is used in order to develop multi-axial constitutive laws for ceramics undergoing martensitic stress-assisted transformation, such as partially stabilized zirconia or $Al_2O_3-ZrO_2$. The internal variable is identified with the volume concentration of transformed particles, and we assume that transformation occurs so that the change in potential energy due to the transformation is maximized.

When the rate of transformation depends on the applied stresses only through the corresponding change in potential energy, it is shown that the inelastic strain rates are along the normal of a stress function in stress space. The constitutive law depends on all three stress invariants. We further discuss specific stress environments such as crack tip fields, the special case of homogeneous transforming particle distribution, and conditions under which normality is not obeyed.

INTRODUCTION

Transformation toughening [1,2] in ceramics may lead to significant enhancement in the fracture toughness of the material. The martensitic transformation of ZrO_2 particles [4] from the tetragonal to the monoclinic crystal structure relieves the stress intensity factor at the tip of a macroscopic crack, thus leading to higher values for the fracture toughness [5]. The same phenomenon has been observed in TRIP (transformation induced plasticity) steels where the transformation is from austenite to martensite [6,7].

In order to determine the amount of toughness enhancement due to the transformation, constitutive laws are necessary that describe how the high stresses that prevail near a crack tip induce the martensitic transformation. Such constitutive laws for partially stabilized zirconia (PSZ) or $Al_2O_3-ZrO_2$ have been published [2,8,9,10]. The constitutive laws account for the volumetric component of the transformation strain, but provide for the effect of the deviatoric component in different ways, such as neglecting it [5,8], or assuming that upon transformation the shear stresses within a transforming particle vanish [9]. All these constitutive models essentially neglect the kinetics of the martensitic transformation from tetragonal to monoclinic of a ZrO_2 particle. Recent experimental work has shown that the kinetics of the martensitic transformation in ZrO_2 particles [11,12] are essentially similar to the kinetics of the martensitic transformation in TRIP steels [6,13]. Furthermore, experimental work has established the significance of the shear component in the transformation of ZrO_2 particles embedded within a matrix of non-transforming ceramic [10].

It is the objective of this paper to show that the analogy in the transformation kinetics between TRIP steels and transformation toughened ceramics leads to constitutive laws valid for ceramics characterized by stress-assisted martensitic transformations. The constitutive law is derived by using the theory of internal variables, and it is described by the functional dependence of the inelastic strain rates on the applied stresses [2,9].

THEORY OF INTERNAL VARIABLES

Consider a macroscopic sample of material of volume V, which is subjected to the uniform macroscopic stresses σ_{ij} with corresponding uniform macroscopic strains ε_{ij}. The set of internal variables ξ_α ($\alpha=1,2,...$), which are collectively denoted by H, characterize the current microstructure of the sample [14,15]. During an increment $d\sigma_{ij}$ in the stresses, energy balance requires that

$$V\varepsilon_{ij}d\sigma_{ij} = d\Psi + \sum_\alpha f_\alpha(\sigma,H)d\xi_\alpha \tag{1}$$

where $\Psi(\sigma,H)$ is a potential which can be identified with internal energy or free energy [14,15], and f_α is the thermodynamic force conjugate to the microstructural rearrangement $d\xi_\alpha$. It is assumed that the set of incremental internal variables $d\xi_\alpha$ describe the microscopic rearrangements at sites in the sample [14].

The increments of the inelastic strains are defined as

$$d^P\varepsilon_{ij} \equiv \varepsilon_{ij}(\sigma,H+dH) - \varepsilon_{ij}(\sigma,H) . \tag{2}$$

It has been shown by Rice [14,15] that if the internal variables ξ_α obey kinetic relations of the form

$$d\xi_\alpha/dt = r_\alpha(f_\alpha,H) , \tag{3}$$

i.e., if the rate $d\xi_\alpha/dt$ of the microstructural rearrangement $d\xi_\alpha$ depends on the stresses σ_{ij} only through the conjugate thermodynamic force f_α, then the rate $d^P\varepsilon_{ij}/dt$ at which $d^P\varepsilon_{ij}$ evolves is given by the normal in stress-space of a potential $\Omega(\sigma,H)$ thus

$$d^P\varepsilon_{ij}/dt = \partial\Omega(\sigma,H)/\partial\sigma_{ij} \tag{4}$$

where the potential Ω is given by [14]

$$\Omega(\sigma,H) = V^{-1} \sum_\alpha \int_0^{f_\alpha(\sigma,H)} r_\alpha(f_\alpha,H)df_\alpha . \tag{5}$$

If necessary, the effect of temperature can also be included [14,15].

APPLICATION TO STRESS-INDUCED MARTENSITIC TRANSFORMATIONS

It has been shown that the kinetics of isothermal martensitic transformations in steels [6,13] and transformation toughened ceramics [11,12] are nucleation controlled. Kinetic experiments in steels have established that the activation energy Q is linearly related to the transformation free energy change per unit volume ΔG by [13]

$$Q = a + b\Delta G \tag{6}$$

where a,b are constants. The rate of transformation can be expressed by

$$dc/dt = h(c) \exp(-Q/RT) \tag{7}$$

where c is the volume concentration of transformed particles, R is the gas constant, T is the temperature and h(c) is a function related to the density of nucleation sites, the nucleation attempt frequency and the volume of the transformed particles [13].

Under the action of an applied stress, and since the martensitic transformation is an invariant plane strain [7,13] the change in free energy ΔG per unit volume of the transformed product is [7,13]

$$\Delta G = -(\tau\gamma_0 + \sigma_n\varepsilon_0) + \Delta G_0 \tag{8}$$

where γ_0, ε_0 are shear and normal strain components associated with the transformation, and τ,σ_n are the corresponding resolved shear and normal stresses due to the applied stresses σ_{ij}. The term ΔG_0 stands for the sum of the change in chemical free energy, contributions due to internal stresses resulting from the transformation, and surface area contributions. ΔG_0 is independent of the applied stresses σ_{ij} [14].

Following the work of Patel and Cohen [16], we assume that the transformation occurs in a direction that minimizes ΔG, or, equivalently, that maximizes the quantity $\sigma_n\varepsilon_0 + \tau\gamma_0$, i.e.,

$$\sigma_n\varepsilon_0 + \tau\gamma_0 = \max = \Lambda . \tag{9}$$

The value of Λ is computed to be

$$\Lambda = \varepsilon_0(\sigma_1+\sigma_2)/2 + [(\sigma_1-\sigma_2)/2](\varepsilon_0^2+\gamma_0^2)^{\frac{1}{2}} \tag{10}$$

where σ_1,σ_2 are the largest and smallest principal stresses, respectively. In terms of the stress invariants, Λ is given by [17]

$$\Lambda = \varepsilon_0\sigma_m + \tau_e\{\cos\phi \ \varepsilon_0/\sqrt{3} + \sin\phi(\gamma_0^2+\varepsilon_0^2)^{\frac{1}{2}}\} \tag{11}$$

where

$$\sigma_m = \sigma_{kk}/3 \tag{11a}$$

$$\tau_e^2 = J = s_{ij}s_{ij}/2 \tag{11b}$$

38

and the angle ϕ is equal to $\omega+\pi/3$ with $0\leq w\leq\pi/3$, and

$$3\omega = \cos^{-1}[(K/2)(3/J)^{3/2}] \qquad (11c)$$

and $s_{ij} = \sigma_{ij} - \sigma_{pp}\delta_{ij}/3$ is the stress deviator with K denoting the determinant of the stress deviator.

The analysis of stress-assisted martensitic transformations is now cast within the framework of the theory of internal variables. We identify one internal variable, namely the volume concentration c of transformed particles; we identify its conjugate thermodynamic force as $(-\Delta G)$, ΔG being the change in free energy upon transformation per unit volume of transformed material. Furthermore, we note that the kinetic expression for dc/dt, Eq. (7), via the use of Eq. (6), depends on the applied stresses only through ΔG. Thus, the potential Ω of Eq. (4) exists, and it can be calculated explicitly via Eq. (5). We finally find that the transformation rate is given by

$$dc/dt = g(c) \exp(\Lambda/\sigma_0) \qquad (12)$$

and that the constitutive law is given by

$$d^p\varepsilon_{ij}/dt = (dc/dt)(\partial\Lambda/\partial\sigma_{ij}) \qquad (13)$$

where the dependence of Λ on the applied stresses is shown in Eqs. (10) or (11), and we have denoted

$$\sigma_0 \equiv RT/b . \qquad (14)$$

The function g(c) of Eq. (12) is related to the function h(c) of Eq. (7) by

$$g(c) = h(c) \exp[-a/RT - \Delta G_0/\sigma_0] . \qquad (15)$$

Eq. (13) establishes the fact that the inelastic strain rates $d^p\varepsilon_{ij}/dt$ are proportional to the normal of the function $\Lambda(\sigma_{ij})$ in stress-space. The shape of Λ is shown in Fig. 1. We observe, from Eq. (1), that it consists of straight-line segments, and that it depends on all three stress invariants, as Eq. (11) shows.

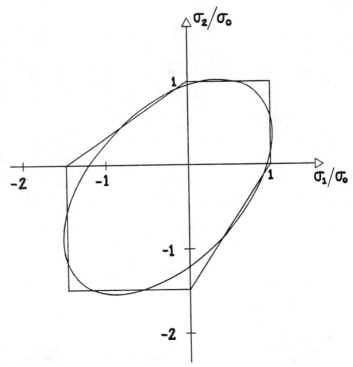

Figure 1:   The surface Λ=constant with ε_0/γ_0=0.2. The straight
line segment is given by Eqs. (10) and (11). The ellipse
is given by Eq. (16).

Simplifications for the Stress Potential

In both steels and ceramics, the martensitic transformation is such
that $\varepsilon_0 \simeq 4\%$ and $\gamma_0 \simeq 20\%$ [6,9]. The fact that γ_0 considerably exceeds ε_0
can be used to simplify the dependence of Λ on the stress invariants.

In Eq. (11) we observe that $\cos(\omega+\pi/3)\varepsilon_0/\sqrt{3}$ is always less than 0.012
in absolute value, whereas in the second term within the brackets
$\sin(\omega+\pi/3)$ lies between 0.87 and 1.0. Approximating the first term by 0,
and the second term by 1, we find that

$$\Lambda \simeq \varepsilon_0\sigma_m + \gamma_0\tau_e \; .\tag{16}$$

When expressed in terms of the principal stresses, the surface Λ=constant
yields an ellipse which is also shown in Fig. 1. Obviously, the expression
in Eq. (16) involves only two stress invariants. Similar linear
combinations between the mean stress σ_m and the equivalent shear stress τ_e
have been discussed in analyzing the effect of shear on transformation
toughening [10,18].

DISCUSSION

The main result of the previous section, namely that the inelastic strain rates are along the normal to a stress potential in stress space, Eq. (13) is a consequence of the kinetic relation given by (7) and of the observation that dc is the only internal variable with $-\Delta G$ as its conjugate thermodynamic force. Due to the lack of quantitative data on the kinetics of martensitic transformations in ZrO_2 particles in ceramics, it is difficult to ascertain the extent of the validity of the assumption that only one internal variable governs the transformation. Indeed, it has been observed that normality is not always satisfied [10]. Clearly, more work is necessary in order to establish quantitatively the kinetics of the transformation of ZrO_2 particles, as well as the inelastic strain rates accompanying transformation. The theory of internal variables, as outlined in the previous section, is expected to provide a convenient method of deriving the constitutive law once the kinetics are established.

The significance of normality, Eq. (13), stems from the observation that the 6 components of $d^p\varepsilon_{ij}/dt$ are known in terms of a single stress potential, namely Λ. Thus, if normality is satisfied, knowledge of Λ alone suffices.

As (13) shows, at each level of σ, the inelastic strain rates are normal to the surface Λ=constant passing through that specific stress point. If all particles at which transformation may occur are identical, then obviously transformation occurs within a time interval dt → 0. Thus, as (13) shows, transformation occurs at a fixed value of Λ, say

$$\Lambda = \text{constant} = \Lambda_0 . \qquad (17)$$

The value of Λ_0 may be found by performing a simple test, such as pure bending or uniaxial compression [10]. Fig. 1 shows the curve given by Eq. (17) which has been calibrated such that inelastic deformation in uniaxial tension occurs at $\sigma=\sigma_0$.

When evaluating $d\Lambda/d\sigma_{ij}$ in Eq. (13) with Λ given by either Eq. (11) or (16), the microscopic strains ε_0 and γ_0 entering the invariant plane strain transformation need not be the same as the macroscopic inelastic strains. Indeed, it has been shown that the ratio of macroscopic shear to dilatation is about one-half of the corresponding ratio of the microscopic quantities [10].

Recent micrographs of transformed partially stabilized zirconia have shown that microscopic shear bands occur [10]. Shear localization initiated by autocatalytically correlated transforming particles has been proposed as the explanation of the manifested macroscopic shear strain [10]. It should prove interesting to examine the relations between the particle and matrix elastic moduli and the stress state dependence of the angle at which the shear bands form. It may be questionable whether or not normality is satisfied when shear bands form. It is expected that more than one internal variable may be necessary in describing the constitutive law in that case.

We conclude by observing that normality of inelastic strain increments to a potential in stress space allows the use of energy methods (such as the J integral [8]) in the calculation of the toughness enhancement due to the transformation [2]. In such a calculation it is not necessary to calculate the detailed stress distribution ahead of a growing crack. Instead, only the residual strain energy within the wake of transformed material left behind the growing crack tip is required [2,8]. Further work is necessary in order to determine the toughness enhancement corresponding to a constitutive law of the form shown in Eqs. (11) and (13).

ACKNOWLEDGEMENTS

The author gratefully acknowledges financial support from the Department of Mechanical Engineering at the University of Rochester, and from the National Science Foundation under Grant MSM-8503984.

REFERENCES

1. A.H. Heuer, F.F. Lange, M.V. Swain, A.G. Evans, J. Am. Ceram. Soc., 69, 181 (1986).
2. A.G. Evans and R.M. Cannon, Acta metall., 34, 761 (1986).
3. P.F. Becher, Acta metall., 34, 1885 (1986).
4. A.G. Evans and A.H. Heuer, J. Am. Ceram. Soc., 63, 241 (1980).
5. R.M. Meeking and A.G. Evans, J. Am. Ceram. Soc., 65, 242 (1982).
6. G.B. Olson and M. Cohen, in Mechanical Properties and Phase Transformations in Engineering Materials, edited by S.D. Antolovich, R.O. Ritchie and W.W. Gerberich (The Metallurgical Society, Warrendale, 1986), p. 367.
7. R.H. Leal, Ph.D. Thesis, M.I.T., 1984.
8. B. Budiansky, J.W. Hutchinson, J.C. Lambropoulos, Int. J. Solids Struct., 19, 337 (1983).
9. J.C. Lambropoulos, Report MECH-55, Harvard Univesity (1984).
10. I.-W. Chen and P.E. Reyes Morel, J. Am. Ceram. Soc., 69, 181 (1986).
11. I.-W. Chen and Y.-H. Chiao, Acta metall., 31, 1627 (1983).
12. I.-W. Chen and Y.-H. Chiao, in Advances in Ceramics, vol. 12, edited by N. Claussen, M. Ruhle and A.H. Heuer (American Ceramic Society, Columbus, 1984), p. 33.
13. G.B. Olson and M. Cohen, Metall. Trans. A, 13, 1907 (1982).
14. J.R. Rice, in Metallurgical Effects at High Strain Rates, edited by R.W. Rohde et al. (Plenum, New York, 1973), p. 93.
15. J.R. Rice, J. Mech. Phys. Solids, 19, 433, 1971.
16. J.R. Patel and M. Cohen, Acta metall., 1, 531 (1953).
17. L.E. Malvern, Introduction to the Mechanics of a Continuous Medium (Prentice Hall, Englewood Cliffs, 1969), p. 91.
18. J.C. Lambropoulos, J. Am. Ceram. Soc., 69, 218 (1986).

THE STRESS INDUCED TRANSFORMATION BY FRACTURE
IN Y$_2$O$_3$ CONTAINING TETRAGONAL ZIRCONIA POLYCRYSTALS

G. KATAGIRI[*], H. ISHIDA[*], A. ISHITANI[*] AND T. MASAKI[**]
[*]Toray Research Center, Inc., Otsu, Shiga, 520 Japan
[**]Technical Development Department, Toray Industries, Inc.,
Otsu, shiga, 520 Japan

ABSTRACT

The stress induced transformation in Y$_2$O$_3$ containing tetra
gonal zirconia polycrystals (Y-TZP) by a three-point bending
test was studied by Raman microprobe. Transformation zone extends to as large as 100μm in the starting side of the fracture
and the transformation zone size becomes smaller in the direction from the starting to the ending side of the fracture. The
transformation zone size has no correlation with fracture
toughness and Y$_2$O$_3$ content. It is suggested that the transformation plasticity may operate in the initial stage of the fracture. The obtained results have remarkable contrast with the
case of the fracture of a pre-indented specimen. The transformation behavior of defect-dominating fracture is also discussed.

INTRODUCTION

Y$_2$O$_3$ containing tetragonal zirconia polycrystals (Y-TZP)
have drawn much interest as a material exhibiting excellent
mechanical properties [1-3]. It is widely accepted that martensitic transformation from tetragonal to monoclinic phase
plays a key role in improving strength and toughness of Y-TZP
[4-8]. Raman microprobe has been proved to be a useful tool for
investigation of the transformation of ZrO$_2$ containing ceramics
[9-12] because of its ability of distinguishing polymorphs with
a μ m order spatial resolution.
In our previous work [12], the transformation behavior of
Y-TZP after indentation and successive fracture was investigated
by Raman microprobe analysis in connection with Y$_2$O$_3$ content.
The profile of transformation zone was determined directly and
the transformation zone size, an important parameter in theoretical studies of transformation toughening, was evaluated.
The results were compared with the theoretical relation between
fracture toughness, K_{Ic}, and the transformation zone size, d,
proposed by McMeeking et al.[7], as follows;

$$K_{Ic} = K_{Ic}{}^{m} + \eta V_f \cdot \Delta V \cdot E\sqrt{d}/(1-\nu) \qquad (1)$$

where $K_{Ic}{}^{m}$ is fracture toughness of a matrix, V_f is a volume
fraction of transformable tetragonal phase, ΔV is the volume
dilation associated with the transformation, E is Young's
Modulus, ν is Poisson's ratio and η is a constant. Expected
linear relationship was experimentally confirmed between toughness increment, $\Delta K_{Ic}(=K_{Ic}-K_{Ic}{}^{m})$ and $V_f\sqrt{d}$.
In the present work, the transformation behavior of Y-TZP
by fracture without prior indentation was examined. The profiles of transformation zone were directly measured not only in
the starting and ending sides of the fracture but also inside of
a test piece by cutting and polishing the specimen. The results

were compared with those of the fracture with prior indentation.

EXPERIMENTAL

The preparation of test specimens was previously described [12]. The composition, the pre-sintering temperature after cold isostatic pressing (CIP) and the hot isostatic pressing (HIP) temperature of each specimen are summarized in Table I. The fracture toughness, K_{Ic}, was evaluated by the indentation micro-fracture method with 30kg load using the equation proposed by Niihara et al.[13]. Fracture toughness of each specimen is also shown in Table I.
A test piece of dimension 3 by 3 by 24 mm was fractured by a three-point bending test under condition of 20 mm span length and 0.5 mm/min loading speed. The fractured test piece was cut into halves and carefully polished as shown in Fig.1.

Table I. Compositions, Sintering Conditions and Fracture Toughness of Y-TZP and Y-PSZ Specimens.

| Composition | | Sintering Temp. | Fracture |
Y_2O_3 (mol%)	Al_2O_3 (mol%)	CIP/HIP (°c)	Toughness K_{Ic} (MPa$\sqrt{m}$)
2.0	0.5	1450/1400	15.6
2.5	0.5	1450/1400	6.0
4.0	0.5	1450/1400	5.1

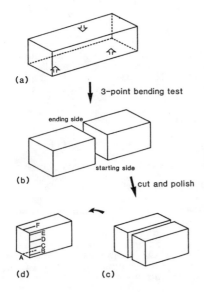

(a)

3-point bending test

ending side

starting side

(b)

cut and polish

(d) (c)

Fig.1. Geometry of measurement for Raman microprobe analysis.
(a) A test piece of a dimension 3 by 3 by 24 mm.
(b) Test pieces after a three-point bending test.
(c) Test pieces cut into halves and polished.
(d) Places of line analysis, A to F.

Raman spectra were measured using a Raman microprobe, MOLE (Jobin Yvon) with 4880 Å line of an Ar ion laser. The optical conditions used were as follows: objective lens x100, numerical aperture 0.9, laser power on the sample 20 mW, the size of the aperture diaphragm 1 mm and monochromator slit widths 600μm. On these optical conditions, 80 % of scattered light was collected from a depth within 9μm [16]. The monoclinic concentration, C_m, was quantitatively estimated by the following equation using peak intensities of monoclinic(m) and tetragonal(t) Raman bands [12],

$$C_m = \frac{1/2 \, (I_m^{181} + I_m^{192})}{k \cdot I_t^{148} + 1/2 \, (I_m^{181} + I_m^{192})} \qquad (2)$$

where k(=2.2) is a correcting factor for the difference of scattering cross-section between monoclinic and tetragonal Raman bands. The superscripts refer to the Raman shift of the characteristic peaks.

The profiles of the transformation zone were obtained by point analyses along the lines denoted A (the starting side of the fracture), B-D (inside of the test piece) and F (the ending side of the fracture) as illustrated in Fig. 1.

RESULTS

The obtained profiles of the transformation zones for 2.0 and 2.5 mol% Y-TZP and 4.0 mol% Y-PSZ are shown in Fig. 2. Measurements were conducted along three different lines in the starting (A) and the ending (F) sides of the fracture. The transformation zone size was evaluated from the profile according to the method defined previously [12] and is indicated by an arrow in Fig. 2. The obtained profiles are complicated, although the changes of the monoclinic concentration are moderate. The transformation zone extends to as large as 100μm in the starting side of the fracture in 2.0 mol% Y-TZP and 4.0 mol% Y-PSZ specimens.

Cracks are observed only in the starting side of the fracture in these specimens as shown in Fig. 3. The monoclinic concentration becomes higher near the cracks in the 4.0 mol% Y-PSZ specimen. The measurements of the profile were conducted along the line where a crack can not be detected by an optical microscope in the 2.0 mol% Y-TZP specimen. On the other hand, the profile is more simple and the transformation zone size is smaller in the 2.5 mol% Y-TZP specimen. A crack can not be observed in both sides of the specimen for this composition.

The transformation zone size generally decreases from the starting to the ending side of the fracture in all the specimens. The profile also becomes more monotonous. In order to examine the relationship between fracture toughness and the transformation zone size, K_{Ic} of each composition is plotted against $V_f \sqrt{d}$ in Fig. 4 similarly as the previous work [12]. No correlation can be observed between K_{Ic} and $V_f \sqrt{d}$, although a linear relationship was observed in the case of the pre-indented fracture [12].

46

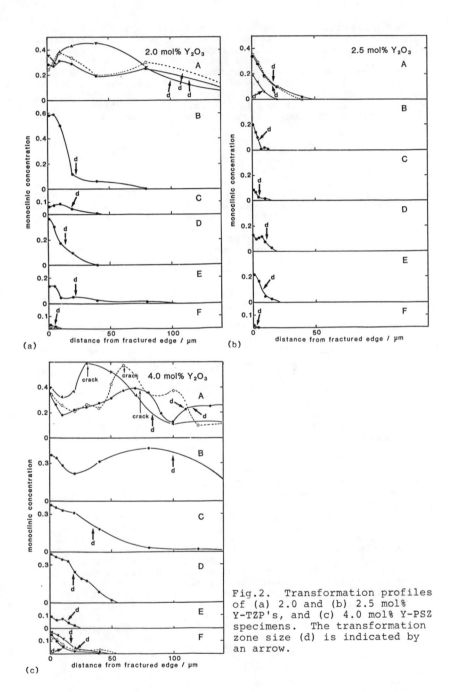

(a)

(b)

(c)

Fig.2. Transformation profiles of (a) 2.0 and (b) 2.5 mol% Y-TZP's, and (c) 4.0 mol% Y-PSZ specimens. The transformation zone size (d) is indicated by an arrow.

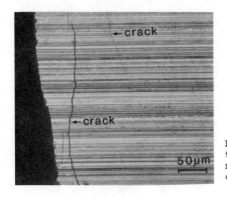

Fig.3. Optical micrograph of the starting side of the 4.0 mol% Y-PSZ specimen. Several cracks are clearly shown.

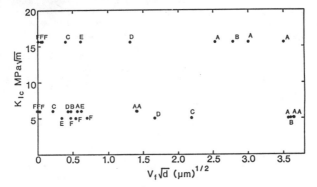

Fig.4. Experimental plot of K_{Ic} against $V_f\sqrt{d}$ in 2.0 and 2.5 mol% Y-TZP's and 4.0 mol% Y-PSZ.

DISCUSSION

The transformation behavior in the fracture without indentation seems to be different from that with indentation in the following respects. First, the transformation zone size is less than 20 μm in 2.0 mol% Y-TZP and is nearly zero in 4.0 mol% Y-PSZ in the case with indentation, whereas it is as large as 100 μm in both 2.0 mol% Y-TZP and 4.0 mol% Y-PSZ. Secondly, the transformation profile is very complicated especially in the starting side of the fracture in the specimens without indentation. Lastly, the transformation zone size or $V_f\sqrt{d}$ of Y-TZP fractured without indentation has no correlation with fracture toughness or Y_2O_3 content.

From these results, it is concluded that the mechanism of the transformation is different between two cases. The transformation by fracture of indented Y-TZP can be considered as stress induced transformation in front of crack tip. The crack-shielding by dilational transformation [7] was proved to be an effective mechanism of increasing toughness, since the linear relationship was observed between ΔK_{Ic} and $V_f\sqrt{d}$ as expected from Eq.(1) [12].

On the other hand, the fracture of the specimens without indentation is considered to proceed as follows. In the initial stage of the fracture, the stress field ranges over large area in the starting side of the fracture and it is relaxed by stress induced transformation at first. This is consistent with the large transformation zone observed in 2.0 mol% Y-TZP and 4.0 mol% Y-PSZ. Several cracks are generated when the transformation reaches to its critical limit. The large monoclinic concentration around the cracks in the 4.0 mol% Y-PSZ specimen may not be due to generation of cracks itself. Cracks may be generated where the applied stress field is fairly large and transformation can not relax the stress field effectively. Because, the complicated profile is also observed in the 2.0 mol% Y-TZP specimen in which the measurements were conducted along the lines which do not across the cracks.

After generation of cracks one of the cracks is enlarged by further application of the stress and the fracture progresses rapidly with growth of the crack. In this stage, the behavior may resemble to that in the specimen with indentation. However, since the stress induced transformation associated with the crack growth is known to be nearly zero in 4.0 mol% Y-PSZ [12], the observed transformation can be considered to be mainly due to the transformation in the initial stage of the fracture.

Chen et al. proposed that transformation plasticity would be operating in highly toughened Mg-PSZ with large transformation zone size [14]. Deformation-induced microcracks were observed in several Mg-PSZ's and it was concluded that microcrack damage must always accompany transformation plasticity.

Transformation plasticity may also operate in the initial stage of the fracture in 2.0 mol% Y-TZP and 4.0 mol% Y-PSZ, since large transformation zone and cracks are observed in the starting side of the fracture.

The different behavior of the transformation profile in the 2.5 mol% Y-TZP specimen compared with the other compositions can be explained by defect-dominating fracture. Small cracks, pores, and inclusions may work as defects. The transformation zone size is small and the profile is monotonous in the 2.5 mol% Y-TZP specimen. No cracks are observed in the starting side of the specimen. These experimental results indicate that the applied stress field concentrates at the defect and then the transformation zone is confined in the vicinity of the defect. Thus, the fracture behavior can be considered to be similar with the case of a pre-indented specimen.

The above explanation is consistent with the other experimental observation [14]. The degree of the transformation was examined by measuring Raman spectrum on fractured surface of many test pieces which are prepared in the same way by pressureless sintering. The degree of the transformation is found to be smaller in the specimens with lower bend strength in which the fracture may be dominated by the defect.

ACKNOWLEDGEMENTS

The authors are grateful to Prof. I-Wei Chen for helpful discussions. Thanks are also due to Y. Ueda for sample preparation and M. Teramura for Raman microprobe analysis.

REFERENCES

1. T. Masaki and K. Kobayashi, in Proceedings of the Japanese Ceramic Society Meeting, 1981, p. 2-3.

2. K. Tsukuma, Y. Kubota and T. Tsukidate, in Advances in Ceramics, vol.12, Science and Technology of Zirconia II, Edited by N. Claussen, M. Ruhle and A.H. Heuer(American Ceramic Soc., Columbus, OH, 1984), p. 382-90.

3. T. Masaki, J. Am. Ceram. Soc. 69(8), 638-40 (1986).

4. A.G. Evans and A.H. Heuer, J. Am. Ceram. Soc. 63(5-6), 241-48 (1980).

5. F.F. Lange, J. Mater. Sci. 17(1), 225-63 (1982).

6. D.B. Marshall, A.G. Evans and M.D. Drory, in Fracture Mechanics of Ceramics, vol.6, Edited by R.C. Bradt, D.P.H. Hasselman, F.F. Lange and A.G. Evans(Plenum, New York, 1983) p. 289-307.

7. R.M. McMeeking and A.G. Evans, J. Am. Ceram. Soc. 65(5), 242-45 (1982).

8. B. Budiansky, J.W. Hutchinson and J.C. Lambropaulos, Int. J. Solids Struct. 19(4), 337-55 (1983).

9. D.R. Clarke and F. Adar, J. Am. Ceram. Soc. 65(6), 284-88 (1982).

10. S. Kudo, J. Mizuno and H. Hasegawa, Yogyo-Kyokai-Shi 94(8), 737-41 (1986).

11. G. Katagiri, H. Ishida, A. Ishitani and T. Masaki, to be published in J. Am. Ceram. Soc..

12. G. Katagiri, H. Ishida, A. Ishitani and T. Masaki, in proceeding of Zirconia '86, Tokyo, Japan, 1986, to be published in Advances in Ceramics.

13. K. Niihara, R. Morena and D.P.H. Hasselman, in Fracture Mechanics of Ceramics, vol.5, Edited by R.C. Bradt, D.P.H. Hasselman, F.F. Lange and A.G. Evans(Plenum, New York, 1983) p. 97-106.

14. I-Wei Chen and P.E. Morel, J. Am. Ceram. Soc. 69(3), 181-88 (1986).

15. G. Katagiri and K. Nakamura, unpublished work.

16. G. Katagiri, unpublished results.

REVERSIBLE TRANSFORMATION PLASTICITY IN UNIAXIAL TENSION-COMPRESSION CYCLING OF Mg-PSZ

K. J. Bowman*, P. E. Reyes-Morel**, and I-W. Chen*
The University of Michigan, Ann Arbor, Michigan 48109-2136

ABSTRACT

Previously, the pressure, temperature and strain rate sensitivities of transformation plasticity have been investigated for monotonic loading of Mg-PSZ. Research in this area has been extended to fully reversed cyclic loading of the type used in plastic strain control fatigue. Cyclic deformation experiments were performed to permit investigation of constitutive behavior under stable deformation conditions at microstrain levels. It was found that cyclic microstrains over a range of temperatures and strain rates were associated with reversible transformation plasticity in the strongly thermally-activated regime. These results are compared to the constitutive relations of transformation plasticity which have been previously developed to explain macrostrain observations.

INTRODUCTION

Among all the transformation toughened ceramics, Mg-PSZ has the highest toughness. This class of zirconia has a coarse grain, two phase microstructure in which the metastable tetragonal phase forms coherent precipitates in a cubic matrix. Stress-assisted transformation plasticity has been studied extensively in this material [e. g. 1-3]. The total volume expansion due to stress-assisted transformation is around 1.5%, which is not reversible below 850K. The dramatic toughening observed in PSZ has been attributed to the macroscopic strains possible from transformation plasticity. In some commercial Mg-PSZ a small volume expansion of 0.15 %, representing roughly 10 % of the metastable phase undergoing a phase transformation is manifested as mechanical hysteresis between 200K and 500K [3-5]. Evidence of some reversible transformation of a similar magnitude has also been reported near the tensile surface in bending at room temperature [6]. It is not clear, however, whether the microstrain transformation affects only a small fraction of the metastable phase or operates by the same mechanism as the irreversible macrostrain transformation which can be driven to exhaustion. This issue is deemed important since it is transformation plasticity in the microstrain regime that is of central importance to fatigue, creep relaxation and slow crack growth at ambient temperatures. The present study represents a first attempt to address this issue.

*Department of Materials Science and Engineering
**Department of Nuclear Engineering

Mat. Res. Soc. Symp. Proc. Vol. 78. ⓒ 1987 Materials Research Society

In this context, a fundamental question to be answered is the interplay between mechanical variables, such as stress, strain, deformation history and the kinetics of forward and reverse transformations manifested as strains. The unique features of stress-assisted martensitic transformation dictate that the stress state effect is important and that the strain partition is anisotropic. In addition, thermal activation is significant as expressed by the strong temperature and strain rate sensitivities. Together, constitutive relations of this kind have provided the crucial elements which allowed us to establish the mechanism of transformation plasticity in the macrostrain regime [see 1]. To pursue a similar approach toward the constitutive relations of microstrain transformation plasticity, we have investigated the response of Mg-PSZ to fully reversed cyclic loading over a range of temperatures and strain rates. The results of this study and its interpretation, in terms of transformation kinetics and transformation plasticity, are reported below.

EXPERIMENTAL

The material studied here is a coarse grain PSZ containing 8 m/o MgO which has previously been characterized at macrostrain levels [1-3]. Fatigue specimens are dog-bone contoured with a 7.5mm diameter cylindrical gauge section. Strain gauges in axial and circumferential directions were used to record strains and an extensometer with an 8mm gauge length was used for feedback to a computer-controlled servohydraulic system. Cyclic loading experiments were conducted in strain control, push-pull fatigue, between preset limits of a symmetric, axial plastic strain amplitude of 5×10^{-5}. Details of this system, and its utilization for cyclic deformation studies will be reported elsewhere. Excellent alignment of the load train was implemented as verified by strain recordings from a triplet of strain gauges spaced annularly at 120° around a test bar.

RESULTS

A typical stress-plastic strain curve is shown in Figure 1, for a plastic strain amplitude of 5×10^{-5}. Because this type of data representation involves subtraction of elastic strain, the portions of the hysteresis loop which are parallel to the stress axis represent elastic behavior. This nearly linear elastic behavior is limited to a stress range of ~50MPa at unloading from the maximum tensile/compressive load. Thus, the 'yield' stress for cyclic microstrain is indeed very low and it is apparently affected by the back stress built up by transformation strain during the previous cycle. Thus, plastic strain is exhibited prior to complete unloading of the specimen from a particular direction of deformation. The plastic portion of the hysteresis loop continues until the strain direction is later reversed at the set plastic strain limits. As verified by the complete absence of an accumulated change in specimen dimensions after many cycles at the same plastic strain amplitude, plastic microstrain of this magnitude is entirely reversible. Once a full hysteresis loop was attained, there was little or no indication of cyclic hardening or softening. Additionally, the loop can be translated upward or downward without a detectable change in shape or size by traversing further into a tensile or compressive region. This suggests that the level of internal stresses opposing the plastic deformation in the reversible microstrain regime is strain dependent and yet reproducible in either straining direction.

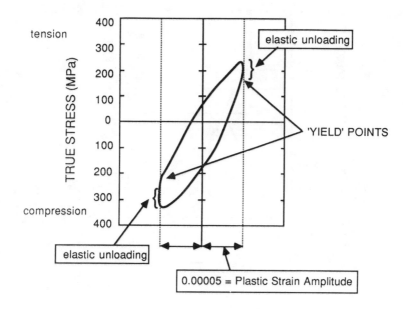

Figure 1: Stress versus plastic strain for second full cycle at room temperature and a strain rate of 4E-5/s.

The above features are all characteristic of hysteresis loops for weakly hardening materials in the microstrain regime. And yet, it is remarkable that the hysteresis loop in Mg-PSZ is asymmetrically disposed. It is noted that the compressive micro-yield stress was ~110MPa for the first cycle, which was started in compression as shown in Figure 2. There was noticeable unloading hysteresis well before zero stress was reached. Nonetheless, the compressive peak stress was over 50MPa greater than the companion tensile peak stress which followed. Subsequent hysteresis loops were nearly identical in the ratio of compressive to tensile peak stresses, such that the peak compressive stress is 1.5 times the peak tensile stress. Lacking any apparent cause of anisotropy which might otherwise be attributed to crystallographic texture, this distinct feature of stress asymmetry in the present material unambiguously signals operation of transformation plasticity, which is a phase transition entailing a volume change. As reviewed elsewhere [1], a similar strength differential effect and a concomitant pressure hardening effect have been documented in the macrostrain regime for transformation plasticity of zirconia ceramics.

54

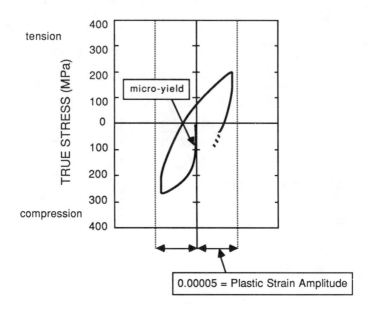

Figure 2: Stress versus plastic strain for the first full cycle at room temperature and a strain rate of 4E-5/s showing initial micro-yield in compression. Note that the plastic strain amplitude was underestimated by ~0.00001 plastic strain on the first cycle.

Two other remarkable features concerning thermal activation were observed for hysteresis loops of Mg-PSZ in the cyclic microstrain regime. The size of the loop under the same plastic strain limits is extremely sensitive to temperature and strain rate. As shown in Figure 3, the stress level decreases significantly at higher temperatures and increases significantly at higher strain rates. The first three loops in Figure 3, with each point representing a recorded data point, show the increase in stress level as strain rate is increased at room temperature. Besides the increase in loop size, the increase in strain rate is shown by the correspondently smaller number of recorded data points at the higher strain rates. The fourth loop, which is actually a composite of hysteresis loops from three consecutive cycles, demonstrates strain rate sensitivity at 383 K. The smaller loops at the lower strain rate, 4×10^{-5}/s, were recorded immediately before and after the larger one for a higher strain rate of 1.6×10^{-4}/s. These results are summarized in Figure 4, wherein the height of the hysteresis loop is plotted against temperature and strain rate. Within a modest temperature range of approximately 100 K, at a constant strain rate, the stress range varies by a factor of two. Likewise, with a modest two decade change of strain rate, the stress level changes by almost 50 %. As reviewed elsewhere, similar, but somewhat weaker temperature and strain rate dependences have been documented in the macrostrain regime for transformation plasticity of Mg-PSZ.

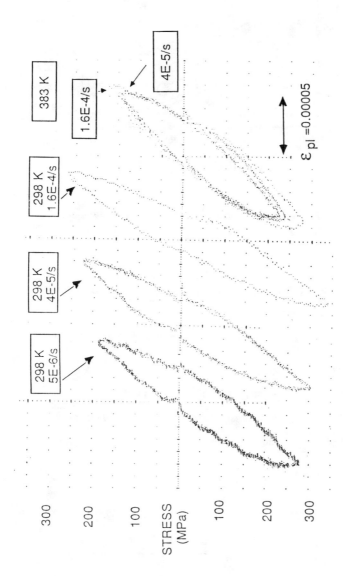

Figure 3: Stress-plastic strain hysteresis loops at a variety of strain rates for both room temperature and an elevated temperature of 383 K. Strain rates and temperatures are indicated by boxes adjacent to loops.

56

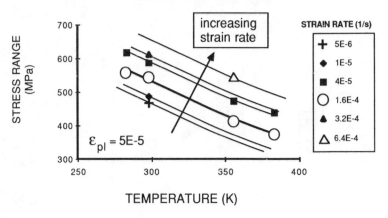

TEMPERATURE AND STRAIN RATE DEPENDENCE OF Mg-PSZ IN CYCLIC MICROSTRAIN

Figure 4: Hysteresis loop height given as stress range versus temperature for several strain rates.

Lastly, to monitor the volumetric strain during cyclic deformation, the plastic strain amplitude was increased to 7.5×10^{-5}. In cycles similar to the one depicted in Figure 1, increased volumes were recorded at the peak compressive load and after subsequent unloading to zero load. Thus, the volumetric strain cycle is apparently in phase with the stress cycle, but with a positive mean value as expected for the stress-assisted tetragonal to monoclinic transformation.

DISCUSSION AND CONCLUSIONS

We have previously formulated a theory which predicts the strength differential based on the mechanical coupling between the applied stress and the transformation [1-3]. In the microstrain regime, it is expected that the most favorably oriented variants will be responsible for most of the transformation observed. The model then predicts that the compressive yield Y_{oc} is related to tensile yield Y_{ot} by

$$Y_{oc}/Y_{ot} = (2 a + 3) / 3 \qquad (1)$$

where a is the pressure sensitivity in compression as expressed by

$$a = 2\hat{e}_T / [\, s_T - (\, \hat{e}_T + e_T \,) / 2].\qquad(2)$$

In the above, e_T is the normal transformation strain across the habit plane of the variant, $\hat{e}_T$ is the volumetric strain, and s_T is the shear strain of the transformation. Using materials parameter $e_T = 0.0253$ (c habit) or -0.0015 (a habit), $\hat{e}_T = 0.0443$ and $s_T = 0.1575$, the prediction gives $a = 0.72$ or 0.68 and a strength differential of 1.48 or 1.43. The latter is very close to the measured value in cyclic microstrain. Note that the above prediction is also expected to be a lower bound. This is because the coupling in the most favorable case has an optimized orientation which tends to compensate somewhat for the inherent disadvantage of compression versus tension due to its negative mean stress. We recall that in the macrostrain regime, the measured pressure sensitivity in compression is 1.25 ± 0.06, which is higher than the above prediction. We have previously attributed this observation to the possibility that macrostrain requires the cooperative transformation of more than one of the variants locally, which are not all likely to be of the optimum orientation for mechanical coupling. Thus, the difference between macrostrain and microstrain transformation plasticity of Mg-PSZ can be rationalized on the basis of variant coupling.

We have also described a theory which considers the kinetics of stress-assisted transformation plasticity [1]. It predicts a strong association between the temperature effect and the strain rate effect. Both are attributed to strong thermal activation which is directly related to the very small activation volume necessary for interface dislocation motion. By comparing the results in the macrostrain regime and in the microstrain regime, this correlation is verified. We may thus conclude that transformation plasticity in the reversible microstrain regime follows the same stress-assisted mechanism as the macroscopic transformation plasticity, but with a somewhat weaker stress-state sensitivity and a somewhat stronger strain rate and temperature sensitivity.

ACKNOWLEDGEMENTS

This research is supported by U. S. National Science Foundation under Grant No. DMR-8609146. K. J. Bowman is supported by the U. S. National Science Foundation under Grant No. DMR-8506705. Partial support for P. E. Reyes-Morel is provided by the Government of Chile.

REFERENCES

1. I-W. Chen and P. E. Reyes-Morel, "Transformation Plasticity and Transformation Toughening in Mg-PSZ and Ce-TZP", this volume.
2. I-W. Chen and P. E. Reyes-Morel, "Implications of Transformation Plasticity in ZrO_2 -Containing Ceramics: I. Shear and Dilatation Effects", J. Am Cer. Soc., **69** [3] 181-189 (1986).
3. P. E. Reyes-Morel, "An Experimental Study of Constitutive Relations of Transformation Plasticity in Zirconia-Based Ceramics", Ph.D. Thesis, Department of Nuclear Engineering, Massachusetts Institute of Technology, (1986).
4. J. A. Excell and M. Marmach, "Reversible Cryogenically Induced Tetragonal to Monoclinic Phase Transformation in Mg-PSZ", Ceramic Bulletin, **65** [10]1404-1407 (1986).
5. S. P. F. Becher, private communication.
6. D. B. Marshall and M. R. James, "Reversible Stress-Induced Martensitic Transformation in ZrO_2", J. Am Cer. Soc., **69** [3] 215-217 (1986).

PART II

Transformation
Plasticity and Toughness

DEFORMATION OF TRANSFORMATION TOUGHENED ZIRCONIA

JAMES LANKFORD
Southwest Research Institute, Department of Materials Sciences, 6220
Culebra Road, San Antonio, TX 78284

ABSTRACT

Recent experimental work on the yield and flow behavior of both single crystal and polycrystalline transformation toughened zirconia is presented. In addition, related work by other researchers is reviewed. The resulting picture is used to assess the relative plastic deformation contributions of phase transformations, dislocation activity, and grain boundary sliding. Particular emphasis is placed on the effects of stabilizer chemistry, grain size, temperature, strain rate, and state of stress. The results are shown to reflect the strong role of shear stresses in selecting, and controlling the operation of, each deformation mode.

INTRODUCTION

While the bend strength and fracture toughness of zirconia have received a great deal of attention during the last few years, relatively little effort has been devoted to the study of general deformation in ZrO_2. This is partly due to the fact that aside from experiments performed at elevated temperature, bend testing does not permit a specimen to deform to a significant extent prior to its failing. However, recent studies [1-8] involving compression, in which both deformation and flaw-nucleated non-interacting microcracks are intrinsically stable, has allowed zirconia crystals and polycrystals to attain more-or-less uniform bulk deformation states. Information derived from such experiments can be very useful, since the latter represent what actually obtains in applications (wear, bearings, confined pressure) for which tensile failure is not dominant. Furthermore, the same basic damage mechanisms probably relate to tensile failure as well, but only within the highly localized regions attending crack tips or fracture origins.

The purpose of this paper is to bring together the results of a three-year study of compressive deformation in zirconia at low and intermediate temperatures, parts of which have been recently published [1-4]. This work, combined with the findings of other investigators regarding deformation at high temperature in both tension and compression, provides at least a partial picture of the plastic flow mechanisms, and their influence on strength, for several major types of zirconia alloys. On the other hand, questions persist as to specific slip systems, apparent interaction between competing deformation/transformation modes, and the origins of certain strain rate-temperature effects; these issues will be noted in the discussion.

EXPERIMENTAL PROCEDURES

The materials investigated in this study represent several generic zirconia alloy configurations; Table I lists the materials, and some of their relevant properties. These are related to one another as follows.

TABLE I

MATERIAL PROPERTIES

Material	Grain Size (μm)	Microstructure	Precipitate Vol. %	K_c (MPa√m)	Wt. % Stabilizer
Mg-PSZ[1]	50	C+T	50	9.5	5 MgO
Mg-FSZ[2]	50	C+T (nontransformable)	15	3.7	3.9 MgO
Y-TZP[3]	0.3	T	--	8.5	9 Y_2O_3
Y-FSZ[4]	single crystal	C	--	1.9	20 Y_2O_3
Y-PSZ[4]	single crystal	C+T	50	6.9	5 Y_2O_3

---

[1]Nilsen TS-Grade PSZ; Nilsen Sintered Products, Ltd, Northcote, Victoria, Australia.

[2]Supplied by M. V. Swain; CSIRO, Melbourne, Australia.

[3]Norton YZ-110 TZP; Norton Co., Northborough, MA, USA.

[4]Ceres Corp., Waltham, MA, USA.

The Mg-PSZ material is a classic coarse-grained, cubic (C) matrix, metastable tetragonal (T) precipitate, transformation toughened zirconia. It clearly contrasts with the coarse-grained Mg-FSZ ceramic, whose tetragonal precipitates are too small to transform under stress, and which therefore should primarily reflect its otherwise cubic microstructure (note the relative values of K_c). Similarly, the behavior of the Y-FSZ single crystal material should reflect the intrinsic deformation of the cubic matrix, absent any grain boundary effects (again note the low value of K_c).

The Y-TZP material, like the Mg-PSZ, is capable of transforming under stress, but possesses extremely fine grains. Both of the latter materials compare with the single crystal Y-PSZ, which represents the intrinsic (no grain boundaries) deformation capability of transformable tetragonal phase in a cubic matrix.

The grain boundaries of the Mg-PSZ and Mg-FSZ were relatively clean [9], although they contain some SiO_2 [10]. The intergranular phase of the Y-TZP material was quite complex, i.e., a continuous film composed of Y_2O_3, SiO_2, Al_2O_3, and possibly ZrO_2 [11].

Cylindrical compression specimens were fabricated from these materials and tested at various temperatures and loading rates; sample preparation and test techniques are described elsewhere [1-4]. The single crystals of Y-FSZ were cut so that the compressive load axis lay along <123>, while Y-PSZ specimens were oriented along either <123> or <100>. Stress-strain plots are presented in terms of engineering stress (σ) versus engineering strain (ϵ); c denotes compression, T tension.

Deformation mechanisms were characterized by means of Nomarski interference microscopy, X-ray diffraction, scanning electron microscopy, and transmission (replica and thin foil) electron microscopy. Space does not permit inclusion of the extensive characterization results per se, most of which are being published elsewhere. Therefore, the following section will emphasize deformation and strength, and the plasticity mechanisms responsible will be introduced via reference.

EXPERIMENTAL RESULTS

Stress-Strain

The deformation behavior of single crystal zirconia is shown in Figure 1. At room temperature, the cubic Y-FSZ material fails in a brittle mode, but as the temperature rises, plastic flow is observed. Microscopic study [3] indicates that this flow is related to extensive wavy slip, which propagates down the specimen in a "front" somewhat like a Luders band; in the case of the ceramic, however, the front is more diffuse than it is for metals, and deformation behind the front is inhomogeneous. The (average) trace of the wavy bands corresponds to slip on {111}, within which shear microcracks eventually form; these probably give rise to the σ-ε serrations, and their coalescence causes ultimate failure.

Deformation of Y-PSZ crystals is extremely sensitive to orientation, as shown in Figure 1(b,c). For the <123> samples, transformation plasticity was observed [3] at 23°C; at higher temperatures, phase transformations still contribute to specimen deformation, but the dominant flow mechanism is now dislocation motion. In this case, slip is very planar [3], and occurs on the single most favored {001}. Shear microcrack formation, and failure, eventually occur within the latter planes.

For crystals with the <100> orientation (Figure 1c), deformation proceeds along a markedly different path. At 23°C, a sudden, stress-dependent, incremental plastic strain occurs prior to brittle failure at a higher stress level. As the temperature is increased, the level required for the strain burst, which has a constant magnitude of 0.0049, decreases. Flow subsequent to the second yield point (load drop) follows a monotonic strain hardening path; its stability and extent (0.15 ultimate strain for T = 700°C) contrast remarkably with the unstable flow, and more limited ultimate strain (0.06) for the <123> orientation at the same temperature.

Based on TEM of surface replicas, it was established [4] that the incremental, pre-macroyield "step" for <100> specimens was caused by the transformation of tetragonal precipitates whose c-axis was parallel to the <100> loading direction. Further, the monotonic hardening, post-yield deformation corresponded to the propagation of a complex Luders front throughout the sample, which failed once the front had encompassed the entire gage section. Thin foil TEM indicated that the deformed region behind the Luders front consisted entirely of two distinct microstructures, i.e., (1) very heavily microtwinned, transformed tetragonal (to monoclinic) precipitates, and (2) a network of highly deformed dislocation cells. The precipitates tended to lie on {110} planes at 45° to the load axis, while the dislocation cell walls appeared to lie in {111} planes. The relative degrees to which these two mechanisms contribute to the hardening is presently unclear.

At still higher temperatures, the deformation situation becomes simpler, and is fairly well understood [5-7]. Fully stabilized <112> cubic crystals compressed at 1400°C deform via a Luders process (Figure

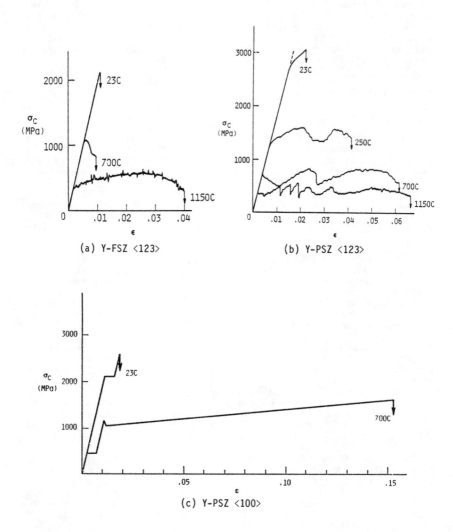

Figure 1. Stress-strain behavior of single crystal zirconia.

2a), controlled by solid-solution hardening of the {100}<110> (predominant) and {111}<110> slip systems. Increasing the Y_2O_3 content of the crystals simply enhances the effectiveness of the solution-hardening, and the yield and flow stresses increase accordingly. If the Y_2O_3 content is reduced, the microstructure at 1400°C consists of a cubic matrix with untransformable T-precipitates; the latter lead to precipitation hardening and serrated yielding, which again occurs on {100}<110>. Within this temperature range, the materials are sufficiently ductile for certain orientations that compression and tension are essentially equivalent, as shown in Figure 2b. However, it is still possible to choose the crystallographic orientation for tension (bending) [12] such that failure is brittle (Figure 2c).

The stress-strain behavior of polycrystalline zirconia both compares and contrasts with that of nominally similar single crystal material, as shown in Figure 3. For example, the essentially cubic (nontransformable) Mg-FSZ material (Figure 3a), is brittle at low temperatures (23°C and 400°C), like the single crystal cubic Y-FSZ. Furthermore, it exhibits a ductility range at 800°C to 1000°C similar to the latter material. However, microscopy shows that the deformation of the polycrystalline zirconia occurs principally via grain boundary sliding and rotation, with concomitant grain boundary cavitation in the glassy grain boundary phase. Grains which cannot slide accommodate the imposed deformation internally by means of wavy slip, presumably on the same slip systems ({111}<110>) [3] involved in the general deformation of the cubic single crystals.

Behavior of the polycrystalline Mg-PSZ (Figure 3b) resembles partially that of the single crystal Y-PSZ, in that it yields and flows at low temperature (23°C) via T → M transformation. However, at 700°C and 1000°C, most of the ductility is accomplished by intergranular creep. Although certain grains do exhibit strain accommodation by transformation plasticity, the strain contribution from this process, as manifested in surface rumpling, clearly declines with increasing temperature. No evidence of dislocation activity could be found either in thin foils or by surface replication.

The declination in transformation strain with increasing temperature is underscored by results obtained in bending [13]. Here (Figure 3d) it can be seen that yielding and flow (limited by failure of the specimen) occurs at 25°C, while at 260°C and 538°C the stress-strain curve is perfectly linear to the point of brittle failure. At 25°C the surface of the sample rumples, indicative of transformation plasticity; not so at the higher temperatures.

The Y-TZP material (Figure 3c), like the Mg-PSZ, yielded and deformed at 23°C solely by means of transformation plasticity. At 400°C and 800°C, however, apparent plasticity is caused principally by intergranular flow, i.e., grain boundary sliding. This sliding is a consequence of the intergranular viscous film; it is interesting to note that the properties of the film are such that grain boundary cavities are observed (via TEM replicas) to form at strain rates on the order of $10^{-4}s^{-1}$ [4]. This means that they formed and coalesced within less than about one minute. At both temperatures, net section deformation appears to take place in shear bands, which at 800°C are virtually superplastic (surrounded by elastic enclaves), and at both temperatures, propagate through the specimen as a "Luders" front. At 400°C, the distribution of cavitated and sliding grain boundaries is quite homogeneous, although incipient shear band formation can be imaged by Nomarski contrast.

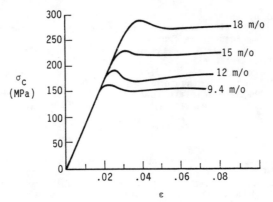

(a) ⟨112⟩ compression axis; T=1400°C; Y_2O_3 content
shown for each curve [5].

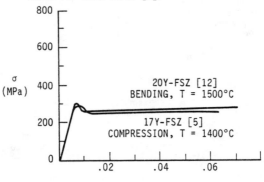

(b) Compression-tension equivalency at elevated temperature.

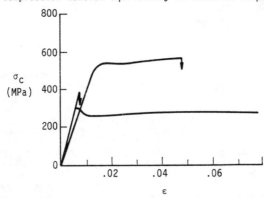

(c) ⟨110⟩ specimen axis; T=1500°C; $\dot{\varepsilon} \approx 10^{-4}s^{-1}$ [12]. Strength/ductility
differences arise from varying crystallographic orientations relative
to bend axis.

Figure 2. Stress-strain behavior of Y-FSZ at elevated temperature.

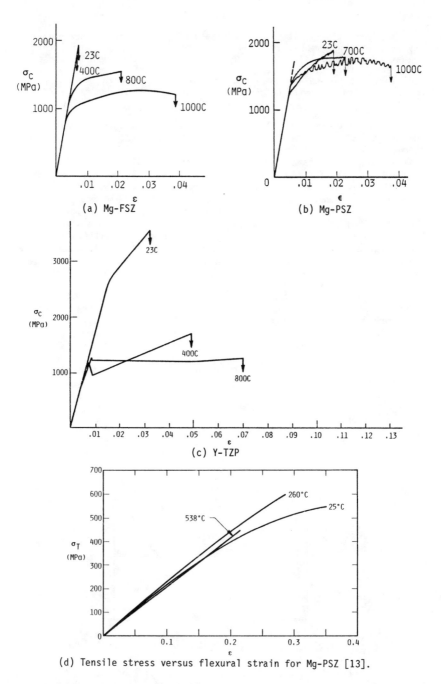

(a) Mg-FSZ

(b) Mg-PSZ

(c) Y-TZP

(d) Tensile stress versus flexural strain for Mg-PSZ [13].

Figure 3. Stress-strain behavior of polycrystalline zirconia.

68

Not surprisingly, it is possible to achieve true (bulk) superplasticity by deforming samples at still higher temperatures, as found recently by Wakai, et al [8,14]. Figure 4, for example, shows the dependence of creep strain upon time and temperature at a low static stress for an Y-TZP [14] very similar to the one used in the present study. It should be noted that strains on the order of unity can be obtained under such conditions.

If the stress-strain results for the Y-TZP are considered in terms of the effect of strain rate at constant temperature, a curious trend emerges. As shown in Figure 5a, specimens tested at 23°c all yield and fail at roughly the same stress level, but the ductility at $\dot{\varepsilon} = 10^3 s^{-1}$ is nearly twice that for the two lower strain rates. At 400°C (Figure 5b), strength rises with strain rate, while the ductility decreases slightly; in addition, the sample tested at $\dot{\varepsilon} = 10^{-5} s^{-1}$ flows following a load drop, while those run at higher strain rates do not. Finally, when the temperature is raised to 800°C, both strength and ductility increase with strain rate, and flow is always preceded by a load drop. It will be recalled that plasticity at $\dot{\varepsilon} = 10^{-5} s^{-1}$, for both T = 400°C and T = 800°C, is caused by homogeneous and inhomogeneous grain boundary sliding, respectively.

This behavior contrasts remarkably with that of similar single crystal material. As shown in Figure 5d, Y-PSZ <100> specimens tested at 700°C deform with increasing difficulty as $\dot{\varepsilon}$ increases; the pre-macroyield stress level increases, as does the stress required for bulk flow, and ductility decreases dramatically (it is nil for $\dot{\varepsilon} = 10^3 s^{-1}$). This represents the inverse strain-rate behavior characteristic of most thermally activated processes, in marked contrast to equivalent results obtained for Y-TZP polycrystals (Figure 5c).

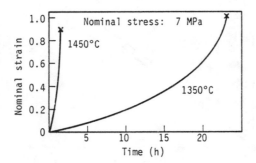

Figure 4. Creep strain versus times at constant load for very fine-grained Y-TZP [14].

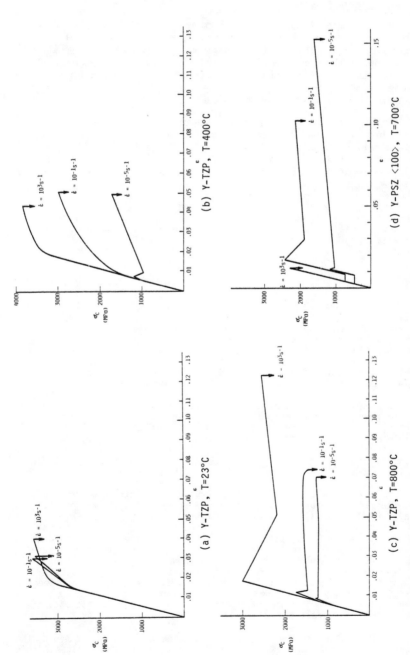

Figure 5. Temperature-strain dependence of compressive yield and flow in polycrystalline Y-TZP and Y-PSZ single crystals.

Strength-Strain Rate

The strength of most monolithic ceramics has been observed to depend upon strain rate as shown for Si_3N_4 in Figure 6. Below $\dot{\varepsilon} \approx 5 \times 10^3 s^{-1}$, σ_c increases slowly with $\dot{\varepsilon}$ at a rate which correlates with the thermal activation of small tensile (even for nominally compressive loading) microcracks. Above this loading rate, the strength undergoes a transition to a regime in which the strength increases very rapidly with $\dot{\varepsilon}$, at a rate essentially independent of temperature [15].

This is not what happens in the case of zirconia single crystals. As shown in Figure 7(a-c), σ_c decreases with $\dot{\varepsilon}$ for T = 23°C, while increasing with strain rate, as expected, at 700°C. The existence of the inverse relationship between σ_c and $\dot{\varepsilon}$ at 23°C is apparently independent of the relative degree of stabilization (Figure 7a,b), as well as orientation (Figure 7b,c).

On the other hand, polycrystalline variants behave in a more conventional fashion (Figure 7(d-f)), aside from the fact that the transition to the inertial regime is not clearly observed at high loading rates. Implications of the foregoing results will be explored in the following section.

DISCUSSION

It is evident that both single crystal and polycrystalline zirconia exhibit plastic deformation at remarkably low homologous temperatures, provided the state of stress is such (in the present case, compressive) that samples do not fail prematurely via flaw-nucleated tensile fracture. This has been noted previously by Swain [16] and by Chen and Reyes Morel [17], who considered only the inelastic deformation inherent in transformation plasticity. The present work extends this concept to dislocation slip and grain boundary sliding deformation mechanisms as well. All of these processes are dominated by shear, and their development in zirconia usually represents the onset of micromechanical instability, as manifested in low hardening rates and multiple load drops. Almost every deformation mechanism discussed in this paper (slip and transformation Luders bands in single crystals; autocatalytic transformation bands in polycrystal grains; superplastic shear bands) initiated from local instabilities. Major questions concern how to model the factors which trigger these several deformation instabilities.

For both single crystals and polycrystals, any strain which occurs at room temperature appears to be generated by phase transformations; essentially no dislocation activity is observed. However, dislocations begin to account for most of the observed single crystal strain at temperatures as low as 250°C (Y-PSZ <123>, Figure 1b), while grain boundary sliding is the principal deformation mechanism in polycrystals at temperatures $>$ 400°C. As noted earlier, transformation strains in Mg-PSZ are negligible at temperatures no higher than 260°C (Figure 3d [13]). While slip and transformations do occur at these temperatures (250-800°C), they seem to serve principally to accommodate the predominant grain boundary (GB) sliding. Slip observed in polycrystals appears to be qualitatively similar to that characteristic of similar single crystals.

Failure of all of the ZrO_2 polycrystalline variants was caused by the coalescence into cracks of cavities nucleated at multiple sites within

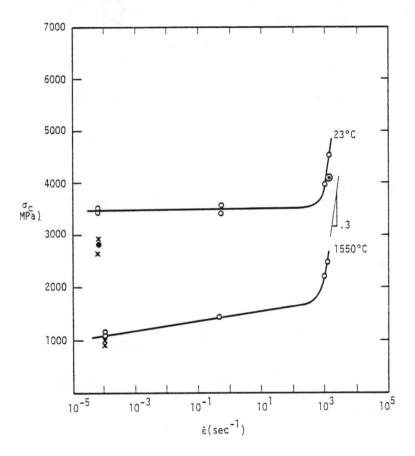

Figure 6. Compressive strength versus strain rate for hot-pressed Si₃N₄.

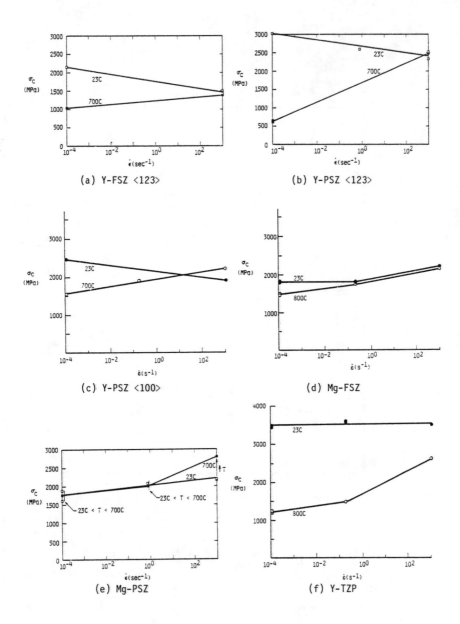

Figure 7. Compressive strength versus strain rate for single crystal and polycrystalline zirconia.

viscous films on GB facets. The cavitation is a direct consequence of the sliding of grain boundaries lying at near 45° angles to the applied stress axis. This is reflected in the macroscopic shear bands observed for Y-PZP at 400°C and 800°C [4]. At 400°C, the GB phase is still sufficiently viscous that only certain boundaries are able to slide at a given stress level; at higher stresses, slightly less favorable boundaries, usually not physically associated with the first to move, are activated. This accounts for the observed hardening and stability of the flow stress curve (Figure 3c). However, at 800°C, many contiguous boundaries can slide at roughly the same flow stress. This produces a low rate of hardening, as "superplastic" bands nucleate and move through the specimen by autocatalysis. It is clear that if the superplastic bands could be prevented, and the GB shear displacements distributed more uniformly, the result would be bulk superplasticity, which has already been achieved at much higher temperatures (1350°C [14]). It is possible that the remarkable increase in both strength and ductility for Y-TZP tested at $\dot{\varepsilon} = 10^3 s^{-1}$ and 800°C (Figure 5c) represents the suppression of instability by the rapid applied loading. This would promote the sliding of many less favorably oriented grain boundaries and inhibit strain localization.

Most of the shear bands which have been observed lie at about 45° to the stress axis. Dalgleish and Evans [16] have attempted to analyze this kind of deformation, and were able to show that to the extent that the macroscopic orientation of the shear bands deviates from 45° relative to the stress axis, the ratio of the GB dilatational strain to the shear strain increases. In the present case, shear bands caused by GB sliding/cavitation invariably lay near 45°, implying that dilatation contributed comparatively little to the macroscopic strain, most of which arose from microscopic GB sliding within the overall plane of the bands.

From Figures 1c and 5d, it is evident that the transformation responsible for the pre-macroyield strain "step" in Y-PSZ <100> crystals is thermally activated. That is, the stress required to initiate the transformation increases with strain rate, and decreases with temperature. It is well known that the transformation supposed to be responsible for inelasticity in these materials is the martensite T → M. The martensite nucleation mechanism responsible for the observed (extreme) temperature-strain rate sensitivity is not known, and merits further study.

It was observed (Figure 7) that at elevated temperatures, the strength of single crystal and polycrystal ZrO_2 increased monotonically. In the case of the former, the increase is due to suppression of thermally activated slip. Similarly, for the large grained polycrystals, the strength was enhanced at higher loading rates by the suppression of viscous GB sliding processes. However, the strength of the fine-grained Y-TZP increased with $\dot{\varepsilon}$ principally because of suppression of instability (shear band) formation.

The basis for the inverse strength-strain rate effect observed for all single crystal variants at 23°C is not yet known. It seems clear, however, that it has nothing to do with transformations, since the effect was present in fully stabilized material (Figure 7a). Since the cracks which cause failure of single crystals have a tendency to lie on certain inclined shear planes, it may be that the sudden imposition of a compressive stress component across the crack face induces such a large friction stress that the cracks cannot shear, and are forced to propagate in a more axial direction. This change in mode might then cause failure at a lower stress level because of easier crack linking (coalescence).

74

ACKNOWLEDGEMENT

The support of the Office of Naval Research under Contract Number N00014-84-C-0123 is gratefully acknowledged.

REFERENCES

1. J. Lankford, J. Am. Ceram. Soc. 66, C212 (1983).

2. J. Lankford, J. Mat. Sci. 20, 53 (1985).

3. J. Lankford, J. Mat. Sci. 21, 1981 (1986).

4. J. Lankford, L. Rabenberg, R. A. Page, J. Am. Ceram. Soc. (submitted).

5. A. Dominguez-Rodriguez, K. P. D. Lagerlof, A. H. Heuer, J. Am. Ceram. Soc. 69, 281 (1986).

6. A. Dominguez-Rodriguez, V. Lanteri, A. H. Heuer, J. Am. Ceram. Soc. 69, 285 (1986).

7. M. L. McCartney, W. T. Donlon, A. H. Heuer, J. Mat. Sci. Lttrs. 15, 1063 (1980).

8. F. Wakai, S. Sakaguchi, K. Kanayama, H. Kato, H. Onishi, presented at the Second International Symposium on Ceramic Materials and Components for Engines, Lubeck-Travemunde, Federal Republic of Germany, 1986 (unpublished).

9. M. V. Swain, J. Mat. Sci. Lttrs. 4, 848 (1985).

10. J. Drennan and R. H. J. Hannink, J. Am. Ceram. Soc. 69, 541 (1986).

11. M. Ruhle, N. Claussen, A. H. Heuer, in Science and Technology of Zirconia II, edited by N. Claussen and A. H. Heuer (The American Ceramic Society, Columbus, OH, 1984), p. 352.

12. R. P. Ingel, D. Lewis, B. A. Bender, R. W. Rice, Com. Am. Ceram. Soc. 65, C150 (1982).

13. D. C. Larsen and J. W. Adams, AFWAL Report No. 10, Contract F33615-79-C-5100, April, 1981.

14. F. Wakai, S. Sakaguchi, Y. Matsuno, Adv. Ceram. Mat. 1, 259 (1986).

15. J. Lankford, Com. Am. Ceram. Soc. 65, C122 (1982).

16. B. J. Dalgleish and A. G. Evans, J. Am. Ceram. Soc. 68, 44 (1985).

Transformation Plasticity and Transformation Toughening in

Mg-PSZ and Ce-TZP

I-WEI CHEN * and P.E. REYES-MOREL**
University of Michigan, Ann Arbor, Michigan 48109-2136

Abstract

The constitutive behavior of transformation plasticity in single phase fine grain Ce-TZP and polyphase coarse grain Mg-PSZ was studied. A material independent pressure sensitivity and a microstructure dependent strain rate sensitivity and temperature sensitivity were found to be associated with the transformation yield stress. Autocatalysis of various extent operates in all cases, forming broad macroscopic shear bands in TZP and fine crystallographic slip bands which terminate at grain boundaries in PSZ. Transformation plasticity is apparently athermal near the M_b temperature in TZP, but not in PSZ except in brittle fracture. A strong crystallographic texture of the transformed phase develops during deformation. These results are analyzed in terms of a shear-dilatant yield criterion and a rate equation which account for stress assistance in martensitic transformation. An elastic-plastic fracture mechanical model is developed to estimate both the shear and dilatation contributions to the transformation plastic work in the crack tip plastic zone. On the basis of approximately equal shear and dilatation contributions to transformation plasticity, the model predicts a transformation zone height which is four times that of the previous model, and a toughness increment which is two times that of the previous model. These predictions are found in good agreement with the reported toughness-zone size relationship. The effect of temperature and chemical stability on toughness is also rationalized.

Introduction

Zirconia containing ceramics have impressive strength and unusual toughness among typically brittle ceramics. The origin of such outstanding properties has been attributed to a phase transformation which imparts apparent plasticity to the crack tip, thus enabling crack shielding and energy dissipation in the process zone. To understand this important phenomenon, we have taken an approach in which we first characterize the constitutive behavior of transformation plasticity, and then model the inelastic process at the crack tip accordingly. Further attention has also been given to elucidate the fundamental mechanisms of martensitic transformations. Reviews of this work have been given periodically in the past two years, covering Mg-PSZ and pure zirconia[1-3]. Our study has since been extended to Ce-TZP and zirconia composites. This expanded scope of the data base now affords us the opportunity to compare the effects of different microstructures on transformation plasticity and to assess the generality of the concepts developed for transformation plasticity. In particular, the comparison of the changing role of thermal activation in transformation plasticity under different circumstances has resulted in new insight into the kinetics of stress assisted transformation. The more accurate description of the constitutive relations of transformation plasticity has also resulted in the successful prediction of toughness enhancement due to shear and dilatational contributions to the plastic work at the crack tip. The present paper is intended to provide a brief mechanistic overview of transformation plasticity and transformation toughening in these materials based on the recent work of Reyes-Morel[4] and other current results from our laboratory.

Microstructural Considerations

The PSZ and TZP studied were two similar Mg-PSZ containing 8 mole % MgO and a

* Department of Materials Science and Engineering
**Department of Nuclear Engineering
Supported by the U.S. National Science Foundation under Grant No.DMR-8609146.
Partial support for P.E. Reyes Morel provided by the Government of Chile.

Ce-TZP containing 12 mole % CeO_2. Schematically, their microstructure can be represented as in Fig. 1. The PSZ has a coarse grain structure in which coherent precipitates of a second phase make up the transformable material, whereas the TZP has a fine grain uniform structure made of the transformable tetragonal phase only. These microstructural features dictate the following considerations which are expected to govern transformation plasticity: (a) in PSZ transformation plasticity must involve both the matrix slip and the precipitate transformation and that the crystallographic restrictions such as slip planes and grain boundary discontinuities must be obeyed by the matrix slip; (b) in TZP transformation plasticity may involve grain transformation only, provided interactions between transforming grains and their neighbors are sufficient to facilitate development of transformation bands which serve to localize deformation into a plane strain mode. These expectations have been fulfilled by experimental observations. As shown in Fig. 2, the deformed Mg-PSZ contains shear bands which are crystallographic in character and terminate at grain boundaries, while the deformed Ce-TZP contains macroscopic shear bands which are similarly oriented with respect to the stress axis in such a way as to allow maximal coupling of the transformation strain and the applied stress. These micrographs can be compared with the schematics illustrated in Fig. 3, in which the mode of deformation in Mg-PSZ is pictured as a cooperative one along easy slip planes as opposed to that in Ce-TZP which is a cooperative one transversing many grains along an optimally stressed direction.

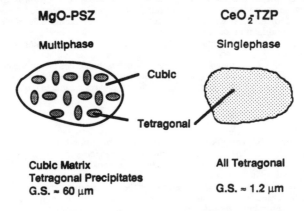

MgO-PSZ **CeO$_2$-TZP**

Multiphase Singlephase

Cubic

Tetragonal

Cubic Matrix All Tetragonal
Tetragonal Precipitates
G.S. ≈ 60 μm G.S. ≈ 1.2 μm

Figure 1. Schematic microstructures of PSZ and TZP materials.

The second important microstructural feature of transformation plasticity concerns the crystallographic bias of the transformation variants which is best revealed by texture analysis of the transformed phase. Unless the transformation strain is isotropically dilatational, as previously envisaged by several investigators[5], transformation strains in transformation plasticity will reflect the stress bias for various micromechanical and macromechanical reasons. An attendant preferred orientation is then expected to accompany such biased macroscopic strains at the microscopic, crystallographic level. Fig. 4 shows the (111) monoclinic reflections which are asymmetrically disposed in different proportions between the doublets depending on the sample orientation with respect to the stress axis. This texture was found in both PSZ and TZP. It was also found that the tetragonal phase in deformed PSZ has no texture, whereas the deformed TZP does. Thus, in Ce-TZP, in addition to transformation, the remaining tetragonal phase itself must have either slip/rotated to accommodate the monoclinic variants or it must have undergone a ferroelastic domain re-orientation to align the tetragonal axis at a certain stage[6]. Microstructurally, partially transformed monoclinic variants and rotated tetragonal variants are expected to coexist within a single grain of Ce-TZP. Because of the difference in tetragonal stability in Mg-PSZ and Ce-TZP, precipitates in PSZ typically either transform in entirety or remain untransformed and undeformed, hence not showing such remnant tetragonal texture. The tetragonal texture can be removed by heating Ce-TZP above the monoclinic-tetragonal reversion temperature, i.e. the A_f temperature at 480°K. In this regard, the present Ce-TZP may be termed a shape memory ceramic.

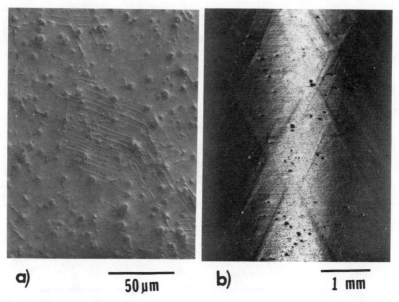

a) ![50 µm] b) ![1 mm]

Figure 2. a)Light micrograph of deformed PSZ, after sectioning and polishing. Shear bands are bounded by grain boundaries. b) Surface light micrograph of Ce-TZP deformed in compression showing macroscopic shear bands.

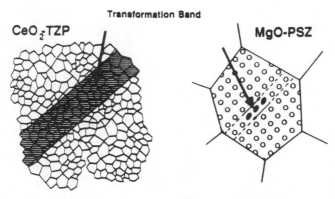

Transformation Band

CeO$_2$-TZP   MgO-PSZ

Figure 3. Schematic representation of the transformation bands in PSZ and TZP.

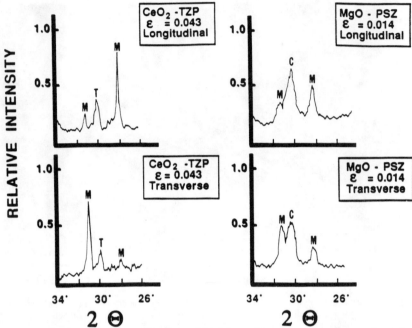

Figure 4. X-ray diffraction of Ce-TZP and Mg-PSZ deformed in compression. A comparison of the (-1,1,1) and (1,1,1) peak heights in differently oriented sections reveals the texture of the monoclinic phase.

The last important microstructural feature of transformation plasticity concerns the microscopic damage incurred during deformation. In general, both materials failed intergranularly in tension and in compression. Thus, like most ceramics, grain boundaries in zirconia are still the weakest members of the structure, despite the overall enhanced toughness. Because transformation involves very substantial shear and dilatational strains, we expect microcracking along grain boundaries to accumulate within transformation bands. This is indeed the case, as shown by the micrographs in Ref. 3. Microcracks formed under stress should be preferentially oriented normal to the maximal principal axis, e.g. axial cracks in compression and transverse ones in tension. This was verified for the coarse grain PSZ deformed in compression[3]. Additional substantiation of this finding was obtained by the measurement of the elastic anisotropy of the deformed PSZ and TZP. As documented elsewhere, the density of microcracks can be determined from such measurements using an anisotropic elasticity model of microcracked solids[4].

Deformation Characteristics

Stress strain curves of PSZ and TZP have been determined in hydraulic compression. In these experiments, a hydrostatic confining pressure is provided in addition to a uniaxial compression. The latter is also referred to as the differential stress. Since, even at a modest pressure, the compression fracture is sufficiently delayed, a full recording of transformation plasticity over the entire range of the transformation fraction can be readily obtained. The measured axial strain is always compressive, whereas the radial strain is always tensile. Alternatively, the volumetric strain is initially compressive in the elastic regime, but subsequently tensile in the plastic regime. A representative set of stress strain curves is shown in Fig. 5. It is noted that the total strain in PSZ is only one third of that in TZP. This is expected, since in the Mg-PSZ studied, tetragonal precipitates in a cubic matrix occupy approximately one-third of the total volume. More remarkably, in TZP there is a nearly perfect plastic regime which is followed by additional plastic deformation at an increasingly higher strain hardening rate. In contrast, the entire range of plastic stress strain curve in PSZ takes on a sigmoidal shape with considerable strain hardening throughout. A sharp yield drop is always observed which precedes the perfect plastic deformation in TZP.

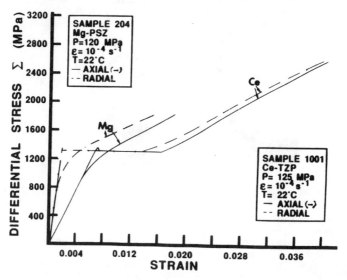

Figure 5. Stress-strain curves for Mg-PSZ and Ce-TZP obtained by triaxial compression under similar conditions: room temperature, 120 MPa, and $10^{-4}s^{-1}$ nominal strain rate.

These characteristics of transformation plasticity are consistent with the microstructural features of the two zirconia as schematically illustrated in Fig. 6. In the dispersed phase zirconia, namely PSZ, each transformation event is ultimately limited to a single grain, although it will typically sweep across all the precipitates on a slip plane. As such, the initiation of each additional transformation event requires an incrementally higher driving force, operating on a certain nucleation site of somewhat lower potency[7]. Strain hardening characteristic of transformation plasticity in the dispersed phase zirconia therefore mirrors the potency distribution of nucleating sites in dispersed zirconia particles[1,7]. Moreover, once the population of transformable precipitates has been greatly depleted, the remaining precipitates tend to have increasingly unfavorable statistics, due to an exhaustion effect, of having even lower potency nucleating sites[7]. The increasingly severe strain hardening at the final stage of transformation plasticity simply reflects worsening nucleation statistics by the demand for a much larger mechanical driving force. The characteristics of the stress strain curve in Mg-PSZ is consistent with the above interpretation if additional provisions for autocatalysis on a slip plane are made. On the other hand, in the homogeneous tetragonal zirconia (Ce-TZP), the transformation of a single variant may trigger transformation of favorably oriented variants in neighboring grains. If this chain of reactions can be sustained over a few grains, the resultant dipole stress field surrounding the transformed region will become a long-range one which can itself trigger further transformations in an autocatalytic way. Thus, macroscopic shear bands form profusely. The above process should be operative if the driving force provided at the outset is sufficient to

80

STRUCTURAL EVOLUTION

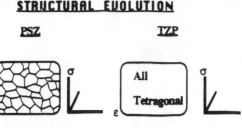

Figure 6.Schematic representation of the microstructural evolution in transformation plasticity.

trigger the chain reaction, which, once initiated, is expected to require a somewhat lower driving force to sustain, hence resulting in a load drop as observed in TZP. The process, however, will be terminated later when shear bands have engulfed a significant fraction of the material, and the initially homogeneous TZP is now partitioned into isolated pockets of transformable phase surrounded by transformed bands. From then on, deformation must become qualitatively similar to that of a dispersed phase zirconia. This accounts for the later stage deformation of TZP.

One important merit of hydraulic compression experiments is the capability of directly probing the stress state dependence of transformation plasticity[3]. This is because of the two independent couplings, in shear and in dilatation, which are provided by uniaxial compression and confining pressure together. Due to the dilatation accompanying the transformation, transformation plasticity under pressure is delayed to a higher flow stress. Thus the measurement of pressure sensitivity of compressive yield provides a means to probe the slope of the yield surface in the hyperspace of shear stress and mean stress. The analysis of the above kind has been performed by Chen and Reyes-Morel for the simple case of a bilinear yield criterion[3]. Experiments on two different types of PSZ and one TZP revealed an apparently similar pressure sensitivity for all three zirconia. As indicated in Fig. 7, the yield stress in compression is elevated by 1.25 ± 0.06 times the confining pressure. Following the analysis of Chen and Reyes-Morel, the critical effective stress $\sigma_e{}^*$ which triggers the transformation is then 0.88 times the critical mean stress $\sigma_m{}^*$ which triggers the transformation. The shear-dilatation yield criterion is written as [3]

$$\sigma_e/\sigma_e{}^* + \sigma_m/\sigma_m{}^* = 1 \qquad (1)$$

in which σ_e and σ_m are the effective and the mean stress, respectively. As discussed in Ref. 3, the effective stress is defined as the deviatoric invariant of the stress tensor, while the mean stress is defined as the hydrostatic invariant of the stress tensor. Their ratio $\sigma_e{}^*/\sigma_m{}^*$ is thus an indicator of the relative weight of the dilatation plasticity versus the shear plasticity associated with transformation in these materials. For the zirconia that we studied, the above two contributions to transformation plasticity are roughly of equal importance.

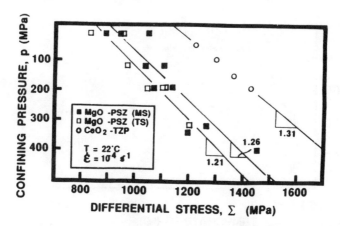

Figure 7. Yield stress versus confining pressure for Mg-PSZ(MS), Mg-PSZ(TS), and Ce-TZP. The slopes of the lines are the pressure sensitivity of transformation plasticity.

The above analysis has been successfully applied to rationalize the large strength differential effect found in tension versus compression[3]. Physically, the origin of the strength differential effect lies in the dilatational component of the transformation strain. The coupling between the applied stress and the transformation strain, however, depends on the relative orientation of the transformation variant. Chen and Reyes-Morel's analysis found that the ratio σ_e^*/σ_m^*, or, alternatively, the pressure sensitivity is larger for an average variant than for the most favorably oriented variant[3]. In the macro-strain regime, the measured ratio σ_e^*/σ_m^* lies in between these two extreme. On the other hand, one should expect that in the microstrain regime, only the most favorably oriented variants will transform and contribute to transformation plasticity. Recent measurements on the strength differential in the microstrain regime in Mg-PSZ subject to fully reversed cyclic loading have confirmed the above prediction[8].

Kinetics of Stress Assisted Transformation Plasticity

A comprehensive study to explore the thermally activated process of transformation plasticity was also undertaken. The main results on PSZ were briefly reported previously[2]. The additional findings on TZP have now shed more light onto this aspect. Essentially, in the autocatalytic, perfect plastic regime, no thermal activation is detected. However, in the strongly strain hardening regime of transformation plasticity in both PSZ and TZP, the strain rate sensitivity is high and the activation volume is low. The latter has been interpreted in terms of a double-kink model for dislocation slip which should be applicable for either interface dislocation motion in martensitic nucleation or for lattice dislocation motion in matrix accommodation[2]. The lack of thermal activation in the perfect plastic regime in which macroscopic shear bands formed was the result of flow localization which tends to propagate at a terminal velocity limited not by the imposed strain rates, but by phonon drag.

An associated aspect in thermally activated transformation plasticity is its temperature dependence. In PSZ, the yield stress has the usual downward trend at higher temperature which is usually associated with matrix slip. On the other hand, in TZP the temperature dependence in the perfect plastic regime is distinctly positive. In particular, it vanishes at a low temperature which coincides with the martensitic burst temperature, i.e. M_b, measured by cryogenic dilatometry. This is shown in Fig. 8 (The spontaneous transformation of Ce-TZP during cooling occurred in an autocatalytic manner, giving rise to an abrupt burst of volume expansion of nearly 1.5 percent. In contrast, in Mg-PSZ, only a relatively gradual volume expansion of 0.1 percent was experienced by cooling below the martensitic start temperature M_s).

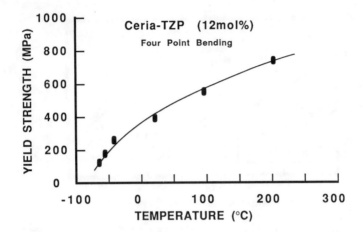

Figure 8. Yield stress versus temperature for Ce-TZP in four point bending. The M_b temperature is at the origin.

As summarized in the previous section, transformation plasticity in zirconia containing ceramics is pressure dependent, suggesting a stress assisted mechanism responsible for transformation plasticity in all such materials under all test conditions. On the other hand, it also manifests both athermal and thermal behavior over a range of temperatures and strain rates, depending on microstructures and the extent of macroscopic autocatalysis. In the past, in the literature on transformation plasticity of steels, the thermally activated regime in which the flow stress decreases with increasing temperature has been customarily associated with the mechanism of strain-induced transformation. In this mechanism, transformation is induced by dislocation slip which activates or even generates nucleation sites for martensitic transformation. Lankford has suggested that this is the case in Mg-PSZ[9]. However, except within a narrow range when the above two mechanisms are both operative and important, the pressure dependence should not be present in strain induced transformation since dislocation slip has at most a very weak pressure dependence. Thus the role of a strain induced mechanism in Mg-PSZ is not likely. In the following, we reconsider the kinetics of stress assisted transformation to demonstrate that both athermal and thermally activated behaviors are possible and it is activation volumes and transformation rates which collectively determine the relative importance of thermal activation in stress assisted transformation plasticity.

Consider a rate equation describing the strain rate due to stress assisted transformation[10],

$$\dot{\varepsilon} = \dot{\varepsilon}_0 \exp \left(- (A - B\Delta G) / kT \right) \qquad (2)$$

In the above, ε is the strain rate of transformation plasticity, $\dot{\varepsilon}_0$ is the maximal strain rate due to martensitic transformation as limited by phonon drag, ΔG is the total driving force, A is an apparent activation energy in martensitic nucleation, B is a constant essentially related to the activation volume of interface dislocations as will become clear later, k is the Boltzmann constant and T is temperature. Here, we have assumed that the strain rate of transformation plasticity is proportional to the rate of transformation. Since the driving force is the sum of the chemical driving force $\Delta G_{chemical}$ and the mechanical driving force W, the latter being dependent on the state of stress, Eqn (2) may be rewritten as

$$W = A/B + (k/B) T \ln (\dot{\varepsilon} / \dot{\varepsilon}_0) - \Delta G_{chemical} \qquad (3)$$

If $\dot{\varepsilon} = \dot{\varepsilon}_o$, the limiting rate of transformation is reached and the second term on the right hand side of Eqn (3) vanishes. This condition applies in the athermal limit in which the required mechanical driving force for stress assisted transformation increases linearly with temperature, assuming a linear relation between $\Delta G_{chemical}$ and temperature. The temperature when $\dot{\varepsilon} = \dot{\varepsilon}_o$ and W = 0 may be identified as the burst temperature of the martensitic transformation M_b. Using these references, we reduce Eqn (3) into

$$W = (k/B) \, T \ln (\dot{\varepsilon} / \dot{\varepsilon}_o) + \Delta S (T - M_b) \qquad (4)$$

where ΔS is the transformation entropy. Since $\dot{\varepsilon} < \dot{\varepsilon}_o$, the first term on the right hand side of Eqn (4) is typically negative. It is now obvious that in Eqn (4) the first term on the right represents thermal softening due to thermal activation whereas the second term represents thermal hardening due to the lowering chemical driving force.

More specifically, the flow stress in proportional loading is related to W by a proportional constant. For uniaxial tension and compression, with a superposed pressure, the proportional constant has been derived previously. We thus write σ = flow stress = βW. The constant β is smaller in tension than in compression and otherwise decreases with increasing mean stress σ_m. With this relation, Eqn (4) becomes

$$\sigma = (\beta k/B) \, T \ln (\dot{\varepsilon} / \dot{\varepsilon}_o) + \beta \Delta S (T - M_b) \qquad (5)$$

Comparing the strain rate dependence embedded in Eqn (5) with the standard definition of activation volume, we readily identify B/β with the activation volume v^* as noted before. The sign of the temperature dependence of the transformation stress in a given material is thus determined by the ratio of v^* to $\ln (\dot{\varepsilon}_o / \dot{\varepsilon})$, since k is a universal constant and ΔS is a material constant.

In Ce-TZP, as shown in Fig. 8, we found σ to increase linearly with T from zero at M_b at a slope which is in excellent agreement with the form given by $\beta \Delta S (T - M_b)$. This may be regarded as the athermal regime. In this regime, ε is very high, limited only by the phonon drag. Autocatalysis proceeds abruptly with a loud sound. The apparent activation volume is very large while the strain rate sensitivity is absent. These features of athermal behavior are entirely consistent with the form of Eqn (5). At higher temperatures, the transformation stress is progressively lowered by an increasing amount from the athermal limit, indicating the increasing importance of thermal activation, although an overall positive temperature dependence is still maintained through the entire range between M_b and A_f. Correspondingly, the intensity of autocatalysis is also reduced. In contrast, the M_b temperature of Mg-PSZ cannot be measured despite some gradual transformation occurring below 223 $^\circ$K in the present material. Neither can the athermal limit be reached in tension and compression using conventional strain rates. Thus, thermal activation is always important and the observed temperature dependence is indeed negative. These various regimes of stress assisted transformation are schematically illustrated in Fig.9. Another consequence of thermal activation as described by Eqn (5) is that a strong positive strain rate sensitivity must be accompanied by a strong negative temperature sensitivity. This is consistent with all the results described thus far. A particularly dramatic example of this correlation were found recently in the microstrain regime during fully reversed cyclic loading of Mg-PSZ as shown in Fig. 10.

Although transformation plasticity in Mg-PSZ in tension and in compression under conventional strain rates seems to be strongly thermally activated, it should be possible at the tip of a propagating crack in zirconia for $\dot{\varepsilon}$ to approach $\dot{\varepsilon}_o$ to regain the athermal limit. This possibility could explain the common observation of decreasing transformation toughening at increasing temperature, in view of the reciprocal relation between toughness increment and transformation stress. We shall return to this point at the end of the next section.

Fracture Toughness

At room temperature, fracture toughness of the PSZ studied here is of the order of 13.5 MPa $m^{1/2}$ and that of the TZP is 10.5 MPa $m^{1/2}$. These values are high compared to other untoughened ceramics. Further studies of microstructural effects and temperature effects on fracture toughness will be reported elsewhere. Below we outline a model which utilizes the yield criterion established experimentally to estimate the crack tip plastic zone and the plastic work dissipation during the steady state propagation of a sharp crack.

84

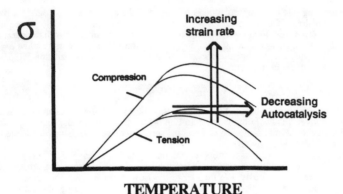

TEMPERATURE

Figure 9. Schematic representation of the effects of strain rate and autocatalysis on the yield stress.

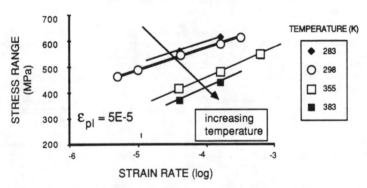

Figure 10. Strain rate and temperature effects on the peak-to-peak stress amplitude in fully reversed cyclic loading in the microstrain regime of Mg-PSZ.

Previous work in this direction has already outlined the shape and the size of the plastic zone and its relationship to the critical transformation stresses[3]. To estimate the plastic work dissipation in the plastic zone, associated with a unit crack extension increment, we use an integral energy balance relation derived by Hutchinson for crack propagation[11]

$$(1-v^2)K^2/E = (1-v^2)K_{tip}^2/E + 2 \int_o^h U(y) \, dy \qquad (6)$$

In the above, K is the applied stress intensity, K_{tip} is the stress intensity at the crack tip and can be equated to the critical stress intensity of the bare matrix in the absence of transformation plasticity. The integral evaluates the residual plastic work over the zone height h, in the far distant wake of the crack, with the integrand as the residual plastic work density given by

$$U = \int [\sigma_e \, d\varepsilon_e + \sigma_m \, d\varepsilon_m] \qquad (7)$$

The latter integral should be evaluated over an appropriate plastic loading-unloading cycle such as the one illustrated in Fig. 11. Assuming normality and perfect plastic deformation in transformation plasticity, for simplicity, the energy integral reduces to $2h\sigma_m*f\hat{e}_t$ in Eqn (6), where f is the volume fraction of the transformable phase and $\hat{e}_t$ is the transformation volume strain. Further substitution of the quadratic proportional relation between h and the mean stress intensity factor taken as $(K+K_{tip})/2$ into Eqn(6) gives the toughness increment

$$\Delta K = K-K_{tip} = X\ Ef\hat{e}_t h^{1/2}/(1-v) \qquad (8)$$

where X is a numerical constant dependent on the ratio of σ_e* and σ_m*. As illustrated in Fig. 12, this ratio approaches 0.2143 in the dilatational limit in agreement with previous studies[12-13] but is twice this value when $\sigma_e* = \sigma_m*$ which is more representative of the yield stress data for both Mg-PSZ and Ce-TZP as found in the last section. In other words, as a result of the roughly equal contributions of shear and dilatation plasticity, the present transformation toughened zirconia must be twice higher and can be regarded as roughly equally attributed to shear and dilatation plasticity. It should be noted that there is a non-additive effect of shear and dilatation in the enlargement of the zone height by a factor of four, as previously demonstrated by us elsewhere[3]. Also plotted in Fig. 12 is the angle between the crack plane and line toward the highest plastic zone extension as defined in Fig. 11. It varies from 90° in the case of the shear criterion (like in metal plasticity) to 61.3° in the case of the dilatation criterion (like in Reference 12-13). For zirconia, at $\sigma_e*=\sigma_m*$, the angle should be around 79°.

The above prediction is found in good agreement with the data plotted in terms of ΔK and f $h^{1/2}$ as in Fig. 13. The implication is that the present yield criterion satisfactorily accounts for the toughness enhancement in PSZ. The case of TZP, however, remains less clear. Ascertaining the fraction of transformable tetragonal phase at the crack tip is of critical importance. It would seem plausible, judging from the currently available experimental evidence and the results of model calculations, that the fraction of the transformable phase must be less than unity to be consistent with the observed toughness in such material. If so, it must then be concluded that only the autocatalytic strains are available for transformation toughening and that the later stage of strain

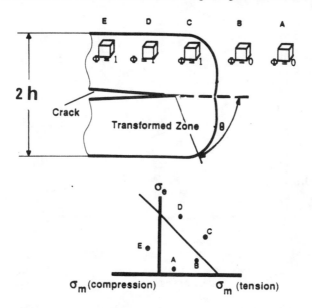

Figure 11. Schematic load-unload path of material elements entering the process zone, with reference to the shear dilatation yield criterion.

hardening is of no consequence. The origin of this has already been pointed out by us previously[2-3]. Essentially, microcracks accompany transformation plasticity in very high concentrations and drastically weaken the material as to prevent it from undergoing further strain hardening in the predominantly triaxially tensile loading at the crack tip. Indeed, for both materials, we found that, while they are flaw tolerant, failure soon occurred once general or localized yielding ensued at a critical stress. Strength and toughness of such materials are thus sensitive to temperatures and stress states, as we previously rationalized[2-3].

In connection to the rate equation of transformation plasticity, it is noted that in both PSZ and TZP toughness decreases with increasing temperature, indicating a positive temperature dependence of the critical transformation stress in both materials. The manifestation of such temperature dependence in PSZ can be rationalized by assuming that a very high strain rate ($\dot{\epsilon} = \dot{\epsilon}_0$ in Eqn (5)) is obtained during crack propagation in brittle materials, hence reversing the temperature dependence which we observed under quasi-static homogeneous loading conditions prevailing in tension and compression tests. In other words, transformation plasticity at the tip of an unstable crack is more athermal than that in smooth test bars.

Despite the strong athermal component of the transformation plasticity experienced at the crack tip, experimental evidence seems to indicate that thermal activation still plays a minor role. Indeed, zone size measurements often revealed a narrowing height of the transformation zones once the crack propagates unstably, suggesting a increasing transformation stress as thermal activation is suppressed. Another manifestation of the thermal activation can be seen by comparing reciprocal toughness of Mg-PSZ ceramics of varying chemical stability at a certain temperature with reciprocal toughness of a certain Mg-PSZ at varying temperature, as shown in Fig. 14. The toughness data plotted here were from Ref. 14 and were previously analyzed in Ref. 2. The line marked "constant T" for data obtained at the same temperature for materials of varying M_s reflects only the contribution of chemical driving force, i.e. the second term on the right hand side of Eqn (5), whereas the line marked "constant M_s" for data obtained over a range of temperature for a material of a given M_s reflects both the athermal and thermal contributions. As expected, the latter is considerably lower that the former.

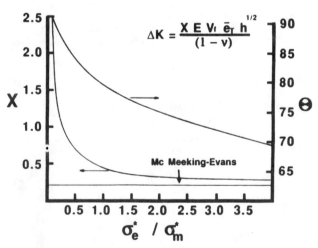

Figure 12. Plot of X versus σ_e^*/σ_m^* ratio. The lower asymptote at $\sigma_e^*/\sigma_m^* = \infty$ coincides with the value given by the McMeeking-Evans model. Also shown is the value of angle θ.

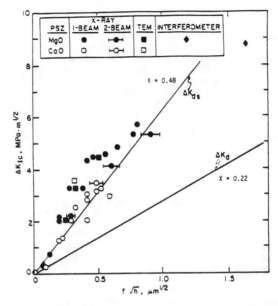

Figure 13. Comparison of the prediction of the shear-dilatation model (upper) and the dilatation model (lower) with experimental data on toughness increments and the zone height.

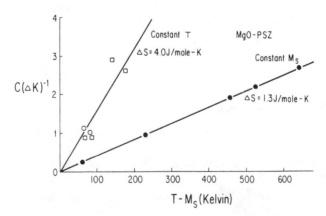

Figure 14. Reciprocal toughness increment scales linearly with M_S temperature and test temperature. The dependence on M_S is much steeper than that on test temperature.

88

Conclusion

A comprehensive study of transformation plasticity in two prototype zirconia ceramics has established the constitutive relations in terms of stress strain behavior and yield criterion over a range of temperatures and strain rates. The constitutive behavior is sensitive to the microstructure because of the varying geometric and crystallographic constraints on the mode of autocatalytic transformation in these ceramics. The nature of transformation plasticity is a stress assisted one, as signaled by its strong stress state dependence under all circumstances. Its energetics and kinetics have been analyzed and are reasonably understood. A theoretical approach has been demonstrated to incorporate the constitutive relations into elastic plastic-fracture mechanical modelling of the crack tip behavior. The model appears to correctly account for the toughness-zone height correlation and the temperature and chemical dependence of fracture toughness.

References

1. I-Wei Chen, "Mechanisms of Transformation and Transformation Plasticity in ZrO_2-Containing Ceramics", in Zirconia Ceramics 4, 55-79, eds S. Somiya and M. Yoshimura, Uchida Rokakuho Pub. Co. Tokyo, Japan(1985).

2. I-Wei Chen, "Implications of Transformation Plasticity on Deformation and Fracture Control of Zirconia-Containing Ceramics", in Zirconia Ceramics 7, 49-63, eds S. Somiya and M. Yoshimura, Uchida Rokakuho Pub. Co. Tokyo, Japan(1986).

3. I-Wei Chen and P.E. Reyes-Morel, "Implications of Transformation Plasticity in ZrO_2-Containing Ceramics: I, Shear and Dilatation Effects", J. Am. Ceram. Soc. **69** [3] 181-189 (1986).

4. P.E. Reyes-Morel, "An Experimental Study of Constitutive Relations of Transformation Plasticity in Zirconia-Based Ceramics", Ph.D. Thesis, Department of Nuclear Engineering, Massachusetts Institute of Technology, June (1986).

5. A.G. Evans and A.H. Heuer, "Review-Transformation Toughening in Ceramics: Martensitic Transformation in Crack-tip Stress Fields"' J. Am. Ceram. Soc., **63** [5-6] 241-248 (1980).

6. A.V. Virkar and R.L.K. Matsumoto, "Ferroelastic Switching as a Toughening Mechanism in Tetragonal Zirconia", J. Am. Ceram. Soc., **69** [10] C-224-226(1986).

7. I-Wei Chen, Y-H. Chiao and K.Tsuzaki, "Statistics of Martensitic Nucleation", Acta Metall., **33** [10] 1847-59 (1985).

8. P.E. Reyes Morel, K.J. Bowman and I-Wei Chen, this volume.

9. J. Lankford,"Plastic Deformation of Partially Stabilized Zirconia", J. Am. Ceram. Soc., **66** [11] C-212-C-213 (1983).

10. G.B. Olson and M.Cohen, "A General Mechanism of Martensitic Nucleation: Part III. Kinetics of Martensitic Nucleation", Metall. Trans., **7A** [12],1915-1923 (1976).

11. J.W. Hutchinson, "On Steady Quasi-static Crack Growth", Harvard University Report, Division of Applied Sciences, DEAP S-8, April (1974).

12. R.M. McMeeking and A.G. Evans, "Mechanics of Transformation- Toughening in Brittle Materials", J. Am. Ceram. Soc., **65** [5] 242-6 (1982).

13. B.Budiansky, J.W.Hutchinson and J.C. Lambropoulos, "Continuum Theory of Dilatant Transformation Toughening in Ceramics", Int. J. Solids Struc., **19** [4] 337-55 (1983).

14. P.F. Becher, M.V. Swain, and M.K. Ferber, "Relation of Transformation Temperature to the Fracture Toughness of Transformation Toughened Ceramics," J.Mater.Sci. **22** [1]   71-84 (1987).

THE TEMPERATURE DEPENDENCE OF YIELD STRESS AND
FRACTURE TOUGHNESS IN UNSTABILIZED ZIRCONIA CRYSTALS

T. W. COYLE*, R. P. INGEL,** AND P. A. WILLGING**
*Ceramics Division, National Bureau of Standards, Gaithersburg, MD 20899
**Ceramics Branch, U.S. Naval Research Laboratory, Washington, DC 20375

ABSTRACT

The flexural strength and the single edge notch beam fracture toughness of undoped ZrO_2 crystals, grown by the skull melting technique, were examined from room temperature to 1400°C. On heating the toughness increased with test temperature to a maximum of 4.0 MPa√m at 1225°C then gradually decreased to 2.6 MPa√m. Upon cooling after a 20 minute hold at 1250°C an increase in toughness to 5 MPa√m was observed at 1200°C; upon cooling to lower temperatures K_{IC} gradually diminished. The load-deflection curves for the flexural strength tests showed marked non-linearity before failure for samples tested on cooling. The temperature dependence of the apparent yield stress suggests that initial yielding occurs by slip above 1200°C but that from 1200°C to 1050°C the observed yielding is due to stress induced tetragonal to monoclinic transformation.

INTRODUCTION

Studies of the mechanical properties of tetragonal ZrO_2 have been limited to ZrO_2-based alloys containing additions such as Y_2O_3 or CeO_2 which permit the tetragonal phase to be retained at room temperature [1]. These materials typically have a fine-grained polycrystalline microstructure and contain appreciable amounts of glassy phase at the grain boundaries [2]. Comparison of the high temperature mechanical behavior of these materials with that of skull melt grown crystals of similar composition [3-6] indicates that these microstructural features dominate the behavior at elevated temperatures. Little information is available concerning the mechanical properties of pure tetragonal or monoclinic ZrO_2 due to the difficulty of preparing unstabilized material suitable for testing. Unstabilized ZrO_2 polycrystalline material extensively microcracks during cooling due to the internal stresses developed by the tetragonal to monoclinic phase transformation. The resulting materials have little or no residual strength.

Large crystals of unstabilized ZrO_2 have been successfully fabricated and provide the opportunity to examine some of the intrinsic properties of pure, bulk ZrO_2. The mechanical behavior of this material at temperatures for which it exists as the tetragonal phase should yield information free from the complications related to the presence of grain boundaries and second phases. The transformation behavior of these materials has been examined versus temperature [7,8] and pressure [9,10]. The narrow transformation hysteresis observed indicates that nucleation of the transformation is relatively easy in these materials compared with the fine-grained polycrystals [11]. These materials therefore also offer the unique opportunity of examining the influence of temperature and applied stress on the tetragonal-monoclinic transformation under circumstances for which the domination of the behavior by nucleation considerations is minimized.

EXPERIMENTAL

Crystals of pure unstabilized ZrO_2[#] were obtained by skull melting [12], a directional solidification process. As reported elsewhere the crystals have been characterized by x-ray diffraction [3,13], dilatometry [7], Raman spectroscopy [8], and TEM analyses [3,14]. Mechanical test specimens were machined from the crystals by standard diamond machining techniques into bars nominally 2mm by 3mm with a 600 grit surface finish. The edges were beveled on a $30\mu m$ diamond lap to avoid failure from edge flaws. Fracture toughness was measured by the single edge notched beam method (SENB) with a notch depth of $\approx 0.5mm$ machined with a 0.15mm diamond bonded cutting blade. Flexure strength and SENB fracture toughness were measured in three point flexure with a span of 25mm as a function of temperature from room temperature (RT) to 1400°C in air. The high temperature experiments were performed in an electrical resistance furnace using high purity Al_2O_3 test fixtures.

The strength and toughness measurements were made under two conditions: on heating to the test temperature and upon cooling to the test temperature from the tetragonal phase field. Samples tested on heating were simply heated to the test temperature, allowed to equilibrate for approximately thirty minutes,then broken. The specimens tested on cooling were first heated to 1250°C (above the finish of the monoclinic to tetragonal transformation) for thirty minutes to insure that the sample was completely transformed to the tetragonal phase, then cooled to the test temperature, allowed to equilibrate, and broken. The apparent yield stress was taken as the stress at which the load-deflection curve became non-linear in the strength tests. On heating these tests were conducted at a straining rate $\delta\epsilon/\delta t \approx 4 \times 10^{-3}$ sec^{-1}. The tests on cooling were conducted at straining rates of $\delta\epsilon/\delta t \approx 4 \times 10^{-3}$ and $\delta\epsilon/\delta t \approx 4 \times 10^{-4}$ sec^{-1}.

RESULTS AND DISCUSSION

Physical and Mechanical Properties

The as-received crystals ranged in size from 25 to 75mm in diameter and up to 120mm in length. The material was opaque white in appearance with relatively flat facets approximately 5 to 10mm wide. On a finer scale (1-3mm) a rounded surface relief was visible on the faces. The physical properties of these materials are discussed in more detail elsewhere [14]. X-ray [3,13], TEM [3], and Raman [8] analyses have shown that the material is completely of monoclinic phase at room temperature. Dilatometry results [7] indicated that during heating the monoclinic to tetragonal (m to t) transformation begins at approximately 1180°C and finishes at approximately 1190°C, and that during cooling the t to m transformation starts at $\approx$1050°C and finishes at $\approx$950°C. The stress free, unconstrained, equilibrium transformation temperature (T_o) for pure bulk ZrO_2 has been estimated as $\approx$1150°C [11].

Examination of thin foils of the crystals at room temperature by TEM [14] has revealed a complex internal structure made up of twin and non-twin related variants. Areas of high dislocation density were observed, and occasionally stacking faults were found. The twins varied in size up to 1000nm in width with lengths exceeding the electron transparent regions of the foils. Microcracks could be observed at the intersections of some variants. The internal structure at temperatures above the m to t transformation is not known. However, the structure of the displacively formed tetragonal phase observed in rapidly cooled Y_2O_3 doped ZrO_2 materials [15,16] suggests that a multi-variant microstructure would exist within these crystals when tetragonal phase as well.

The temperature dependence of the fracture strength and the fracture toughness are shown in Figs. 1 and 2, respectively. It is important to

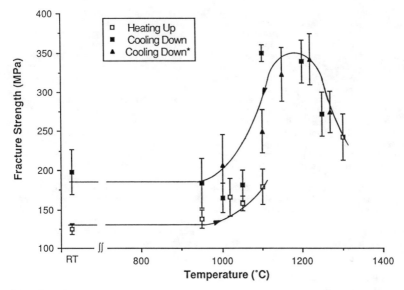

Fig. 1. Fracture strength versus test temperature for samples measured on heating to the test temperature and on cooling from 1250°C to the test temperature at strain rate $\delta\epsilon/\delta t \approx 4 \times 10^{-3}$ sec.$^{-1}$. Error bars represent 95% confidence interval. (* signifies slower strain rate of $\delta\epsilon/\delta t \approx 4 \times 10^{-4}$ sec^{-1}).

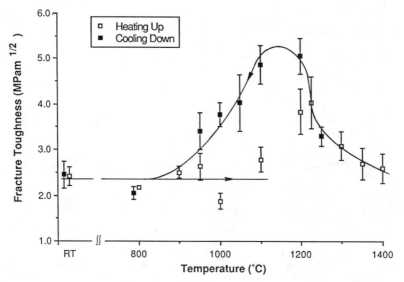

Fig. 2. Fracture toughness (SENB) versus test temperature for samples measured on heating to the test temperature and on cooling from 1250°C to the test temperature. Error bars represent 95% confidence interval.

keep in mind the phase composition of the specimen at the time of testing while considering these results. During heating the material is completely of monoclinic phase at temperatures up to ≈1180°C. The strength and toughness are essentially constant over this temperature range, although a slight increase in the average values is suggested above ≈1000°C. Completely linear load-displacement curves were obtained at room temperature and at 800°C. However, slight deviations from linearity were detected for some specimens tested above 1000°C.

The material has completely transformed to tetragonal phase upon heating above 1200°C, and remains tetragonal during cooling to ≈1050°C. Thus, tests performed under these conditions reflect the behavior of undoped, single phase tetragonal specimens. Strength and toughness both increased as the testing temperature was decreased from 1400°C, reaching a maximum at 1100°-1200°C. Strength and toughness then gradually decreased with decreasing test temperature until 800°C, at which point the t to m transformation had been completed and the toughness had returned to the value measured for monoclinic ZrO2 on heating. The increased fracture strength at 800°C and room temperature of the specimens tested after cooling from the high temperature annealing treatment can be attributed to a change in the flaw population, since the fracture toughness was not affected by the heat treatment. Similar strength increases were obtained by careful polishing of specimens which were not heat treated.

Temperature Dependence of Yield Stress

The apparent yield stress observed in many of the flexural strength tests provides insights into the mechanisms contributing to the strength and toughness behavior. The value of the yield stress as determined from the load-displacement curve is approximate, particularly at low applied loads, due to the effects of the load train used for the high temperature tests. In Fig. 3 the range of the observed yield stresses at each temperature where yield was detected has been plotted versus temperature for the two straining rates employed. The yield stress increased with decreasing test temperature from 1300°C to 1150°-1200°C indicative of thermally activated plastic flow. Shown for comparison in Fig. 3 are data for yield stress as determined by compressive loading for skull melt grown crystals of ZrO_2 containing additions of 2.8 mole % Y_2O_3 and 12 mole % Y_2O_3 tested at temperatures from 25°C to 1150°C [4], and 9.4 and 18 mole % Y_2O3 tested at 1400°C [17]. The 2.8 mole % material is in the two phase, tetragonal plus cubic phase field at the test temperature but still yielded by the formation of planar slip bands which extended through the entire specimen [4]. The structure of the other three compositions is single phase cubic. Interpretation of the present results for the region above 1150°-1200°C as yield controlled by the thermally activated motion of dislocations within the tetragonal structure is consistent with the limited available high temperature plasticity data.

Below 1150°-1200°C the yield stress decreased rapidly with decreasing temperature, approaching zero at 1050°C, the temperature at which the t to m transformation begins spontaneously during cooling. This temperature dependence is that expected for a stress induced martensitic transformation [18]. The large transformation strains involved in the transformation lead to a large contribution to the total driving force from the work done by the applied stresses on the transformation strains. As the magnitude of the chemical free energy change decreases with increasing temperature, an increase in the contribution of the work term to the total driving force is required to cause transformation. The temperature dependence of the critical stress required to nucleate the transformation, identified here as the temperature dependence of the yield stress, is therefore dominated by the temperature dependence of the chemical free energy change for the transformation. Thus, the observed yield behavior in this temperature range can be attributed to the stress induced transformation of tetragonal phase to monoclinic phase under load. This type of transition from a yield stress governed by the criterion for stress assisted nucleation of the martensitic transformation to a yield stress which represents the initiation of true plastic flow is known in metallic systems [19], and has been discussed in detail for the case of TRIP steels [20]. Employing the terminology of ref. 19, the temperature at which spontaneous transformation begins on cooling with no applied load ($\approx$1050°C) is the familiar M_s temperature. The temperature above which initial yield occurs by plastic flow ($\approx$1150°C) is referred to as M_s^σ.

The source of the observed deformation strain below M_s^σ is expected to be primarily the result of the applied stress biasing the formation of the monoclinic variants during transformation, with some contribution due to plastic accommodation of the transformation strains. To the extent that plastic accommodation is absent, the observed strains should be completely recovered following the completion of another thermal transformation cycle in the absence of applied stress, the shape memory effect [21]. Specimens loaded to fifty per cent of the room temperature fracture load during cooling from 1250°C exhibited substantial deformation at room temperature. Heating these specimens above the finish of the m to t transformation, then cooling with no applied load resulted in significant recovery of the previous deformation. Similar behavior has recently been reported for MgO-partially stabilized ZrO_2 (Mg-PSZ) in which the transforming phase is present as dispersed precipitates within cubic grains [22].

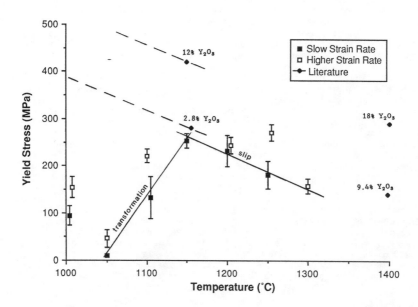

Fig. 3. Apparent yield stress versus temperature for samples measured on cooling to the test temperature. Error bars represent the range of experimental results at each temperature. Literature results shown at 1400°C from [17] and at 1150°C and below from [4].

Temperature Dependence of Fracture Toughness

The strength and fracture toughness results show the same trends with temperature above 800°C, indicating that the mechanisms responsible for the hysteresis in the measured fracture toughness are controlling the fracture behavior. The peak in toughness observed in the range 1100°-1200°C during cooling occurs just above the M_s temperature. An increasing fracture toughness with decreasing temperature, reaching a peak at a temperature slightly above Ms, is the temperature dependence observed experimentally and predicted theoretically for the transformation toughening mechanism [1]. In conjunction with the temperature dependence of the yield stress which indicates that stress induced transformation is responsible for the initial yielding for temperatures up to 1150°-1200°C, this behavior establishes transformation toughening as the mechanism responsible for the increased K_{IC} in the region of the toughness maximum. Below M_s the decrease in K_{IC} can be attributed to a reduction in the volume fraction of tetragonal material available to transform in the zone surrounding the crack tip and/or to an increase in the width of the transformed zone to a point that it is no longer small with respect to the specimen size. The width of the zone would approach infinity (i.e. the sample boundaries) as the stress required to cause nucleation of the transformation approached zero [1].

At temperatures above the peak toughness (>1200°C) the temperature dependence is consistent with that expected for transformation toughening, as stated above. The comparison of fracture surface features of specimens broken under the various testing conditions has given qualitative support to the suggestion that transformation contributes to the observed fracture toughness above M_s^σ [14]. The fracture surface of specimens broken at room temperature, which were composed entirely of monoclinic phase at the time of testing, was extremely rough. Specimens broken at 1100° on heating (monoclinic under test conditions) and at 1100°C on cooling (entirely tetragonal at time of testing) were nearly identical in character to the room temperature specimens. The specimen broken at 1100°C on cooling was clearly in a regime where transformation about the advancing crack tip would contribute to the observed K_{IC}, and result in the crack propagating through a structure which was composed essentially of monoclinic phase. The fracture surfaces of specimens broken at 1200°C and above gradually became more planar. Cleavage-like fracture steps became increasingly common and of larger scale, while on a fine scale the roughness decreased. The fracture surface features thus suggest that the extent of transformation about the propagating crack gradually diminished with increasing test temperature over the range of 1100° to 1300°. At the highest temperatures examined the fracture toughness of the tetragonal phase material had decreased towards that measured for monoclinic ZrO_2 (2.0-2.5MPa√m) and cubic ZrO_2 (1.5-2.5MPa√m) [5].

The apparent effectiveness of the transformation toughening mechanism in these materials at temperatures near or even above the equilibrium transformation temperature (T_o), in contrast to typical transformation toughened systems, is related to the relative ease of nucleation in the present samples. Since the total driving force necessary to induce transformation is much smaller, comparatively low levels of applied stress can be effective in driving the t to m transformation even at temperatures near T_o where the chemical free energy change is small [11,14].

SUMMARY AND CONCLUSIONS

The temperature dependence of the fracture toughness and flexural strength of unstabilized zirconia crystals exhibit a hysteresis which is related to the well known hysteresis in transformation temperature. Yielding behavior was observed before failure in strength tests conducted on specimens which were of tetragonal phase under the test conditions. The temperature dependence of the observed yield stress has indicated that up to the M_s^{σ} temperature of $\approx 1150°$ initial yield occurs by the stress assisted nucleation of the t to m transformation. From the temperature dependence of the fracture toughness and the fracture surface morphologies it was concluded that the stress induced transformation was responsible for the enhanced toughness observed over the temperature range extending from the M_s temperature to 1250°-1300°C.

CERES Corporation, North Billerica, MA., 01862

REFERENCES

1   A.G. Evans and R.M. Cannon, "Toughening of Brittle Solids by Martensitic Transformations", Acta Metall. 34 [5] 761-800 (1986).

2   M. Ruhle, N. Claussen and A.H. Heuer, "Microstructural Studies of Y_2O_3-Containing Tetragonal ZrO_2 Polycrystals (Y-TZP)", pp. 352-370 in Advances in Ceramics, Vol. 12, Science and Technology of Zirconia II. Edited by N. Claussen, M. Ruhle, and A.H. Heuer, American Ceramic Society, Columbus, OH, 1984.

3   R.P. Ingel, Ph.D. Thesis, Catholic University of America, Washington, DC, 1982, Univ. Microfilms Int. #8302474.

4   J. Lankford, "Deformation and Fracture of Yttria-Stabilized Zirconia Single Crystals", J. Mat. Sci., 21 1981-1989 (1986).

5   R.P. Ingel, D. Lewis, B.A. Bender, and R.W. Rice, "Temperature Dependence of Strength and Fracture Toughness of ZrO_2 Single Crystals", J. Am. Ceram. Soc. 65 [9] C150-C152 (1982).

6   J. Lankford, L. Rabenberg, and R.A. Page, "Deformation ande Damage Mechanisms in Yttria-Stabilized Zirconia", submitted to J. Am. Ceram. Soc.

7   J.W. Adams, H.H. Nakamura, R.P. Ingel and R.W. Rice, "Thermal Expansion Behavior of Single Crystal Zirconia", J. Am. Ceram. Soc. 68 [9] C228-C231 (1985).

8   C.H. Perry, D.-W. Liu and R.P. Ingel, "Phase Characterization of Partially Stabilized Zirconia by Raman Spectroscopy", J. Am. Ceram. Soc. 68 [68] C184-C187 (1985).

9   S. Block, J.A.H. deJornada and G.J. Piermarini, "Pressure-Temperature Phase Diagram of Zirconia", J. Am. Ceram. Soc. 68 [9] 497-499 (1985).

10 B. Alzyab, C.H. Perry and R.P.Ingel, "High Pressure Phase Transitions in Zirconia and Yttria Doped Zirconia", Submitted to J. Am. Ceram. Soc. (1986).

11 T.W. Coyle, "Transformation Toughening and the Martensitic Transformation in ZrO_2", Sc.D. Thesis, Massachusetts Institute of Technology, Cambridge, MA, February, 1985.

12   V.I Aleksandrov, V.V.Osiko, A.M. Prokhorov and V.M. Tatarintsev, "Synthesis and Crystal Growth of Refractory Materials by RF Melting in a Cold Container", Current Topics in Materials, Vol. 1, Chapter 6, E. Kaldis, ed. North-Holland Publishing Co. (1978).

13   R.P.Ingel and D. Lewis, "Lattice Parameters and Density for Y_2O_3-Stabilized ZrO_2", J. Am. Ceram. Soc. 1986.

14   R.P.Ingel, P.A. Willging, B.A.Bender and T.W. Coyle, "The Physical and Thermo-Mechanical Properties of Monoclinic Single Crystals", to be published in the Proceedings of the Third International Conference on the Science and Technology of Zirconia, Tokyo, Japan, Sept., 1986.

15   B.A. Bender, unpublished work

16   V. Lanteri, A.H. Heuer, and T.E. Mitchell, "Tetragonal Phase in the System ZrO_2-Y_2O3", pp. 118-130 in Advances in Ceramics, Vol. 12, Science and Technology of Zirconia II.   Edited by N. Claussen, M. Ruhle, and A.H. Heuer, American Ceramic Society, Columbus, OH, 1984.

17   A. Dominguez-Rodriguez, K.P.D. Lagerloff, and A.H. Heuer, "Plastic Deformation and Solid-Solution Hardening of Y_2O_3-Stabilized ZrO2", J. Am. Ceram. Soc. $\underline{69}$ [3] 282-284 (1986).

18   J.R. Patel and M. Cohen, "Criterion for the Action of Applied Stress in the Martensitic Transformation", Acta Metall. $\underline{1}$ 531-538 (1953).

19   G.B. Olson and M. Cohen, "A Mechanism for the Strain-Induced Nucleation of Martensitic Transformations", J. Less-Common Metals, $\underline{28}$ 107 (1972).

20   G.B. Olson, "Transformation Plasticity and the Stability of Plastic Flow", Pp. 391-423 in proceedings of the ASM Materials Science Seminar "Deformation, Processing, and Structure", St. Louis, MO, 1982, American Society for Metals, 1983.

21   G.B. Olsen and M. Cohen, "Thermal-Elastic Behavior in Martensitic Transformations", Scripta Metall. $\underline{9}$ 1247-1254 (1975).

22   M.V. Swain, "Shape Memory Behavior in Partially Stabilized Zirconia Ceramics", Nature $\underline{322}$ 234-236 (1986).

STRENGTH AND TOUGHNESS OF MG-PSZ AND Y-TZP MATERIALS AT CRYOGENIC TEMPERATURES

S. VEITCH[*], M. MARMACH[*] AND M.V. SWAIN[**]
[*] Nilcra Ceramics Pty. Ltd., Northcote, Victoria
[**] CSIRO Division of Materials Science, Clayton, Victoria 3168 Australia

ABSTRACT

The influence of temperature between R.T. and liquid nitrogen (-196°C) on the strength and toughness of three grades of Mg-PSZ and a 3 mol% Y-TZP material have been investigated. The toughness of the Mg-PSZ materials passes through a maximum at the M_s temperature, whereas the strength increases monotonically with decreasing temperature regardless of M_s. The Y-TZP material exhibits increasing toughness with decreasing temperature but the strength passes through a maximum at - 80°C. The results are discussed in terms of transformation and R-curve limited strength of transformation toughened ceramics. For Mg-PSZ materials it is suggested that the critical stress to initiate a non reversible transformation on the tensile surface is responsible for microcracking and ensueing R-curve development.

INTRODUCTION

The influence of temperature on the mechanical properties has received a significant amount of attention because of the technological and basic importance of such data. The majority of studies have been at and above room temperature because of simplicity and applicability[1]. In all these observations the strength and toughness generally decrease with increasing temperature, in agreement with thermodynamical considerations. Very few studies have examined in detail the influence of cooling below room temperature (R.T.) on the mechanical properties. Two studies of this topic are by Lange [2] on Y-TZP (plus Al_2O_3) and Mg-PSZ materials by Becher et al [3]. In both these studies the major emphasis was placed upon the fracture toughness which passes through a maximum at M_s, the temperature for the onset of the t- to m- ZrO_2 transformation, below which it decreased. The present study examines in more detail the influence of temperature between liquid nitrogen (-196°C) and R.T. on the flexural strength and toughness.

The mechanical properties of Mg-PSZ may be modified by various heat treatments as discussed by Hannink and Swain [4]. In the present circumstances the grain size and precipitate size of the Mg-PSZ is kept constant with the mechanical properties modified by sub-eutectoid heat treatment. This occurs by the development of a magnesia enriched δ phase ($Mg_2Zr_5O_{12}$) about the precipitates which leads to a strained interface [5]. The results for Mg-PSZ are compared with a Y-TZP (3 mol % Y_2O_3) material sintered at 1500°C with a grain size of ~ 0.5 μm. Martensite transformations undergoing heterogeneous nucleation, as pointed out by Olsen [6], may be stress or strain assisted, the latter being initiated at new nucleation sites by plastic strain. This concept derived for TRIP steels has been extended by Chen and Reyes-Morel [7,8], for Mg-PSZ and Ce-TZP materials. The major difference between the PSZ and TZP materials is that the transformable tetragonal phase in the PSZ is separated by a non transformable cubic matrix phase. Cooperative or correlated transformation of adjacent t-ZrO_2 precipitates requires that severe deformation strains are accommodated in the cubic matrix between these precipitates. Whereas for the TZP materials events in one grain may readily trigger transformation in an adjacent grain, Hannink et al [9].

Mat. Res. Soc. Symp. Proc. Vol. 78. ⓒ 1987 Materials Research Society

Becher et al [3] recently summarized the thermodynamic stability of t-ZrO_2 precipitates and grains. The chemical free energy change associated with the transformation may be written as

$$\Delta G^{t \to m} = (T - M_s)\Delta S \tag{1}$$

where ΔS is the entropy difference between the two phases. When $T = M_s$, equation (1) implies $\Delta G^{t \to m}$ is zero and the transformation initiates spontaneously. At temperatures above M_s the t-ZrO_2 phase is stable and some external energy must be supplied to transform it. The toughness increment due to transformation toughening is given by

$$\Delta K^T = A \, e^T V_f E \sqrt{h} \tag{2}$$

where A is a constant dependent upon zone shape about the crack tip, e^T is the volume dilation associated with the transformation, V_f the volume fraction of transformed zone, E the elastic modulus and h is the zone height normal to the crack plane. Combining equations (1) and (2) and relating the zone size to the crack tip stress intensity factor K_o, via the small scale yielding criteria and to the critical stress to trigger the transformation σ_c^T, Becher et al [3] arrive at the following relationship between ΔK^T and M_s^c temperature, namely

$$\Delta K^T = \Omega A(e^T)^2 V_f E K_o / \Delta S (T - M_s) \tag{3}$$

where Ω = constant.

The determination of strength for brittle materials is traditionally given by the simple relationship

$$\sigma_f = K_c / Y \sqrt{c} \tag{4}$$

where K_c is the critical stress intensity facator, Y is a geometric factor and c is the radius of a defect or flaw. However for transformation toughened materials that exhibit R-curve behaviour the situation is more complex. In these materials steady state K_c ($K_o + \Delta K^T$) is only achieved after crack extension of a few times the transformed zone size. Two suggestions have been proposed for the limitation of strength in such materials, namely the existence of the critical stress to trigger the t$\to$m-ZrO_2 transformation σ_c^T, at which stress ductility is observed leading to nucleation of microcracks and ultimately catastrophic failure, and the slope of the R-curve [10,11]. Evidence in support of transformation related ductility has been observed for Mg-PSZ and Ce-TZP materials in tension and compression [8-11]. In the case of the Mg-PSZ associated with the onset of non reversible transformation and in-elastic deformation, microcracks were often observed to initiate at grain boundaries . These microcracks continue to grow in a stable manner (R-curve) upon further loading.

Steinbrech and Heuer [12] have attempted to predict the strength of transformation toughened materials on the basis of the R-curve. The critical condition for the onset of unstable fracture occurs when the slope of the applied stress intensity factor $K_I = Y\sigma\sqrt{c}$ equals the slope of the crack resistance curve K_R, that is

$$K_I = K_R \text{ and } \frac{dK_I}{dc} = \frac{dK_R}{dc} \tag{5}$$

The major limitation of the above approach is the paucity of experimental data on R-curves for transformation toughened materials.

EXPERIMENTAL

The material examined was a Mg-PSZ alloy of a nominal 9.4 mol% MgO composition and sub-eutectoid heat treated to three grades namely AF, MS and TS, the latter two are comparable to readily available commercial grades[+]. Details of the phase assemblage as determined by X-ray diffraction and optical microscopy upon grinding or polishing, precipitate size and M_s temperature using a thermal expansion technique are listed in Table 1. All specimens had linear thermal expansion curves on heating above R.T. The Y-TZP material contained nominally 3 mol% Y_2O_3 and was sintered at 1500°C for 2 hours and had a grain size of 0.5 μm. No M_s temperature was observed upon cooling to -196°C.

Sufficient material was removed to eliminate any trace of a residual surface compressive layer generated by grinding. Each bar was given a slight chamfer of the tensile face edges to minimise premature failure from edge initiated cracking.

Fracture toughness tests were performed using the modified indentation technique (MIT) [13] and the applied moment double cantilever beam (AMDCB) [14]. The former were carried out on polished bars pre-indented with a Vickers indent at 20 Kgf at R.T. The AMDCB specimens had dimensions of 3 x 20 x 100mm with a 1.5mm deep guide notch. These specimens were annealed at 800°C for 1 hour to minimise the presence of residual stresses associated with specimen preparation.

Optical micrographs using interference contrast techniques were taken of the polished surfaces after fracture to examine the nature of any deformation generated.

RESULTS

a. Toughness

Some problems were experienced in the determination of toughness with considerable variability between methods. The MIT technique indicated only modest changes in K_c with test temperature for Mg-PSZ and was considered to be an invalid test for this material. The AMDCB were in better agreement with previous estimates of K_c by Becher et al [3]. The latter were felt to be more reliable in that 5-40mm of stable crack extension often occured during the determination of K_c. The AMDCB results for Mg-PSZ and MIT 3Y-TZP values are compared in Figure 1. The Mg-PSZ values all pass through a maximum in the vicinity of the M_s temperature whereas the TZP values increase with decreasing temperature.

b. Strength

Measurements of polished and ground strengths of all the Mg-PSZ alloys showed a monotonic increase with decreasing temperature. The ground surface values were always higher than the polished surface results except for the Y-TZP materials. For the TZP materials the ground and polished results passed through a maxima at -80° and -140°C respectively, although the scatter with these values is somewhat higher than for the Mg-PSZ materials. The results are compared in Figure 2.

c. Optical Micrographs

Optical micrographs of the uniform tensile region of Mg-PSZ bars fractured at various temperatures are shown in Figures 3-5.

[+] Nilcra Ceramics Ltd, Northcote Vic Australia

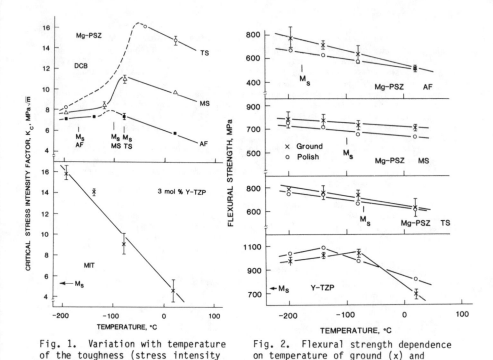

Fig. 1. Variation with temperature
of the toughness (stress intensity
factor) of Mg-PSZ.

Fig. 2. Flexural strength dependence
on temperature of ground (x) and
polished (o) surfaces of Mg-PSZ

The AF material, Figure 3, exhibited no surface deformation features
when fractured at R.T. At -80°C there is a vague hint of some deformation
with the occasional evidence of micro-crack formation. The uniform stress
region at -140°C has a roughened appearance which was almost non-existent in
the vicinity of the outer roller supports (zero stress) and increases in
density toward the central region of the bar. At -196°C there is a much
more significant alteration in surface morphology. A very dense dimpled
rumpling has developed along the entire specimen that becomes significantly
more dense as one progresses toward the constant stress region of the bar.
There was no evidence of microcracking on the tensile face.

In contrast, the MS material, Figure 4, with increased metastability of
the tetragonal phase exhibited microcracking and rumpling at R.T. At -80°C,
almost the M_s temperature, surface microcracking and some rumpling occurred
but 'softer' and more undulating than at R.T. A significant change occurs at
-140°C (below M_s). The microcracking and rumpling have disappeared to be
replaced by surface dimpling within the tensile zone which increased at
-196°C.

The TS grade material, Figure 5, exhibited the most significant degree
of microcracking and rumpling. At -80°C (below M_s) the extent of
microcracking and rumpling was reduced. At -140°C a texture perpendicular
to the tensile stress developed with no obvious microcracking. Once again
at -196°C the extent of this texture became more pronounced. The density of
these band like features was a maximum in the region of uniform stress with
little evidence for such behaviour in the zero stress region outside the
outer rollers.

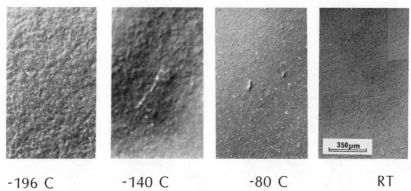

-196 C   -140 C   -80 C   RT

Fig. 3.   Optical micrographs of the tensile surface of AF Mg-PSZ.

-196 C   -140 C   -80 C   RT

Fig. 4.   Optical micrographs of the tensile surface of MS - Mg-PSZ.

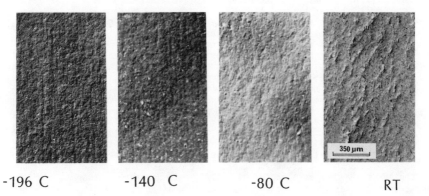

-196 C   -140 C   -80 C   RT

Fig. 5.   Optical micrographs of the tensile surface of TS Mg-PSZ.

DISCUSSION

The toughness results shown in Figure 1 for Mg-PSZ alloys follow the results of Becher et al [3] and the predictions of equation (3). The toughness appears to maximise near the M_s temperature and thereafter to decrease. The maximum values of K_c ~16 MPa $\sqrt{m}$ for the TS material at -40°C is very comparable with the results of Becher et al for a similar material. No attempt was made to monitor the transformed zone size so quantitative agreement with equation (3) is not possible. The maximum achievable value of K_c appears to decrease with decreasing M_s temperature. The Y-TZP results exhibit a monotonic increase in K_c with decreasing temperature because the M_s temperature of this material is below -196°C. The maximum observed value of ~15 MPa$\sqrt{m}$ is comparable to values obtained by Becher (private communication) on similar materials.

The limited results obtained to date and the AMDCB technique do not readily lend themselves to determine whether an R-curve exists for the materials in question. Indirect evidence for an R-curve may be obtained from the optical micrographs in Figures 3-5. The presence of stable microcracks that have undergone extension is evidence for R-curve behaviour and used to measure the R-curve [10,11]. The observations in Figures 3-5 indicate that the AF material exhibits minimal microcracking with the possible exception being at -80°C. The MS material shows microcracking at RT and -80°C with a similar trend for the TS material. For both the MS and TS materials the extent of microcracking and hence of the R-curve decreases at -80°C.

Approximate construction of the R-curve is possible from the observations of the microcracks on the micrographs mentioned above. The stress at fracture is known and the K_c of the microcrack may be determined from equation (4). Further data may be obtained by considering microcracks generated outside the inner rollers assuming a linear decrease in stress with distance between the inner and outer loading rollers. From such considerations it is possible to plot the R-curves shown in Figure 6. This approach may well exaggerate K_c by overestimation of the outer fibre tensile stress in the event of stress induced transformation in the outer tensile region [10,11]. The results shown in Figure 6 reveal a dramatic reduction in the extent of crack extension with decreasing temperature for the MS and TS material. From AMDCB K_c results it might have been anticipated that for the MS materials the extent of the R-curve would have been more extensive at -80°C. The slope of the results in Figure 6 is almost approximately the same suggesting that this technique for determination of the R-curve may have some intrinsic errors.

For the materials and temperatures considered in Figure 6 it appears that an R-curve may have determined the strength but the significant reduction of crack extension with decreasing temperature is contrary to expectation on the basis of the proposed R-curve dependence of strength. As outlined by various authors [10,11] the toughness increment and range of the R-curve scale with the transformed zone size. Steinbrech and Heuer [12] have attempted to quantify this further (section 1) and it would have been anticipated that as the temperature approaches M_s the R-curve slope should decrease because the toughness (and the zone size) becomes very high. The present results are not in accord with this interpretation.

The results at first glance also do not appear to support the notion of a transformation limited strength as proposed by Swain [15] and as extended by Swain and Rose [11]. This proposition would have the strength decreasing as the temperature approaches M_s because the critical stress σ_c to trigger the transformation must approach zero. The temperature dependence of strength, regardless of heat treatment, is unaffected by the M_s temperature as shown in Figure 2.

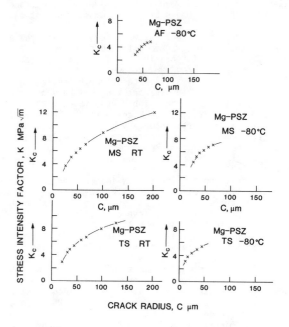

Fig. 6. Approximate R-curves constructed from microcracks on the tensile surface of Mg-PSZ samples at various temperatures.

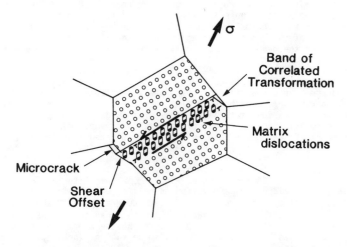

Fig. 7. Schematic diagram of correlated transformation and the formation of microcracks and dislocations in the matrix.

However the latter approach modified by the proposal of Olsen [6] for TRIP steels and extended by Chen and Reyes-Morel [7] provides a framework to interpret the Mg-PSZ results. The basis of the argument now runs as follows.

At room temperature the TS and to a lesser extent MS displayed a readily observable non-linear loading curve with the onset of the non-linearity occuring at 300 MPa and 550 MPa for the TS and MS respectively. For all materials upon cooling to -80°C and below no such behaviour was readily observed in the loading curves, despite strength testing in some instances at the M_s temperature. When microcracking has been observed at room temperature in TS material it is invariably associated with correlated transformation in adjacent grains [19]. Such correlated transformation is equivalent to a shear offset at a grain boundary interface which may assist in nucleating a microcrack Figure 7. This situation is clearly seen about Vickers hardness impressions in tough Mg-PSZ materials [10,15,16]. Chen and Reyes Morel [7] have suggested that the behaviour of matrix with precipitates is to act as a "liquid-drop".

An alternative consideration for the stress induced correlated transformation based upon Olson's outline for TRIP steels is as follows. To achieve a shear offset the correlated transformation requires that the matrix by some means accommodates the very severe localised stresses about the precipitates. It is considered that such large shear offsets in a correlated band of transformation may exceed the ability of the cubic matrix to accommodate such strains in an elastic manner. That is, dislocations are generated within the cubic matrix. Transmission electron microscopy observations of correlated transformation bands in PSZ by Hannink and Swain [16,18] did not reveal any dislocations. However, this matrix material within the transformed band was severely strained making such observations extremely difficult. If dislocations were developed in Mg-PSZ then slip would most likely take place upon the slip systems {100} <110>. Such plastic deformation of the matrix is not required for the random (non correlated) transformation upon cooling through the M_s temperature of these materials.

As suggested by Olson [6], for strain induced transformation in TRIP steels, the critical stress σ_c to trigger the correlated transformation is temperature dependent because for strain induced transformation (in metals) the yield stress is thermally activated. In the Mg-PSZ materials the metastability of the precipitates and pre-strain within the matrix is dependent upon the development of the δ phase due to sub-eutectoid ageing. Thus the stress to cause plastic deformation of the matrix is influenced by the δ phase development. That is, a plot of this superimposed stress to initiate deformation with temperature will be a series of near parallel lines dependent upon the state of the δ phase development. An attempt from the limited data generated here to plot such a series of curves is shown in Figure 8 for the various materials. Some of this data was obtained by noting the position for the onset of deformation (rumpling) or microcracking between the inner and outer loading points of polished bars loaded to failure. Also plotted are hardness data for Y-cubic stabilised zirconia (Y-CSZ) single crystals from Hannink and Swain [16]. Other observations obtained by cooling specimens with M_s above R.T. (350°) under load through M_s [20] and labelled OA are shown on Figure 8. These materials displayed considerable correlated transformation as manifested by deforming up 0.2 - 0.3% inelastic strain under stresses of 150-200 MPa at 300°C and less.

The Y-TZP results are somewhat more limited than the Mg-PSZ. Unlike the latter the strength appears to pass through a weak maximum whereas the toughness continues to increase. Optical observations (not included here) revealed no evidence for microcracking even at -196°C. The absence of microcrack suggests that R-curve behaviour may not be the strength limiting aspect but rather transformation related deformation. It might have been anticipated that with further cooling the stress for the onset of the transformation related deformation would decrease. However the strength measurements are almost constant with temperature below -80°C. If the transformation related deformation was somehow thermally activated, due to the necessity of dislocation motion, then the strength would have been anticipated to slightly increase with decreasing temperature. Further measurements of Y-TZP strengths with differing stabilizer contents are required to investigate and resolve this issue.

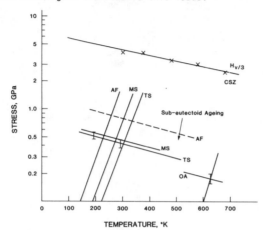

Fig 8.   Proposed influence of temperatures on the critical stress to initiate correlated transformation for the various Mg-PSZ materials. See text for details of how data points obtained.

CONCLUSIONS

The major conclusions of this study are:
(i)   The toughness of Mg-PSZ materials passes through a maximum at the M_s temperature.
(ii)   The apparent flexural strength of Mg-PSZ materials increased with decreasing temperature regardless of the M_s temperature, whereas the strength of Y-TZP materials passed through a maximum at -80°C.
(iii) Microcracking and correlated transformation due to the application of stress generally took place above M_s at a stress determined by the heat treatment condition.
(iv)   No evidence of R-curve behaviour was observed for the Mg-PSZ materials below M_s.
(v)   The initial stress for the onset of correlated transformation of precipitates in Mg-PSZ appears to be dependent upon the extent of δ-phase development upon sub-eutectoid ageing as does the M_s temperature.
(vi)   It is suggested that correlated transformation requires plastic deformation of the matrix and is thermally activated.

REFERENCES

1. D.J. Green, R.H.J. Hannink, M.V. Swain and D.B. Marshall, "Transformation Toughened Ceramics" CRC to be published.
2. F.F. Lange, J. Mater. Sci. 17, 255 (1982).
3. P.F. Becher, M.V. Swain and M. Ferber, J. Mater. Sci., 22, [1] 1987.

4. R.H.J. Hannink and M.V. Swain, J. Aust. Ceram. Soc., 18, 53 (1982).
5. M.V. Swain, R.H.J. Hannink and J. Drennan, "Ceramic Microstructures - 86", Proc. of Conf. on Ceramic Metal Interfaces, Berkeley, July 28-31, 1986 to be published.
6. G.B. Olsen, ASM Materials Science Seminar, "Deformation, Processing and Structure", 391-425 (1984).
7. I-W. Chen and P. Reyes-Morel, J. Am. Ceram. Soc., 69, 181 (1986).
8. P. Reyes-Morel, Ph.D Thesis MIT 1986.
9. R.H.J. Hannink, B. Muddle and M.V. Swain, Proc. 12th Aust. Ceram. Soc., August 1986 P. 145-152.
10. D.B. Marshall, J. Am. Ceram. Soc., 69, 173-180 (1986).
11. M.V. Swain and L.R.F. Rose, J. Am. Ceram. Soc., 69, 511-518 (1986).
12. R. Steinbrech and A.H. Heuer to be published.
13. R.F. Cook and B.R. Lawn, J. Am. Ceram. Soc. 66, C-200 (1983).
14. S.W. Freiman, D.R. Melville and R.W. Mast, J. Mater. Sci. 8, 1527 (1973).
15. M.V. Swain, Acta Metall. 33, 2083-88 (1985).
16. R.H.J. Hannink and M.V. Swain, "Deformation of Ceramic Materials II". p. 695 Plenum Press (1984).
17. M.V. Swain, Nature 322, 234-6 (1986).
18. M.V. Swain and R.H.J. Hannink, p.225, Adv. in Ceramics Vol. 12, Science and Technology of Zirconia II. Edts N. Claussen, M. Rhule and A.H. Heuer.
19. D.B. Marshall and M.V. Swain to be published.
20. M.V. Swain, to be published in Science and Technology of Zirconia Ceramics III. Tokyo, Sept. 1986.

TABLE I. Data of Materials Evaluated

| Material | Monoclinic Content % | | | | Grain Size, μm | Precipitate Size, μm | M_s Temp. °C |
	Ground	Polished	G. Boundary	Matrix			
9 $^m/_o$ Mg-PSZ $_{AF}$	13	2	1	1	50	0.2	-160°
MS	22	4	2	2	50	0.2	-100°
TS	32	15	3	12	50	0.2	- 70°
3 $^m/_o$ Y-TZP	7	0	-	-	0.5	-	< - 196°

CRACK PROPAGATION IN MG-PSZ

M.J. Readey[*], A.H. Heuer[*], and R.W. Steinbrech[**]
[*]Case Western Reserve University, Cleveland, Ohio 44106
[**]University Dortmund, Dortmund, Federal Republic of Germany

ABSTRACT

Certain coarse-grained Mg-PSZ's exhibit marked R-curve behavior, e.g., the crack resistance increases from ~5 to ~15 MPam$^{1/2}$ with the first 500 microns of crack extension. These high toughness materials also show unusually-good flaw tolerance.

Crack propagation and crack path development in Mg-PSZ were investigated during stable crack growth using a modified DCB specimen geometry. Crack extension was monitored with a traveling microscope equipped with a video recording system. PSZ's with various heat-treatments were tested to determine the effect of microstructure on R-curve behavior.

Cracks were found to propagate discontinuously and to arrest. Smaller microcracks would develop around the arrested crack, and link together to form the macrocrack which led to eventual failure.

INTRODUCTION

During the last decade, transformation-toughened MgO partially-stabilized ZrO$_2$, Mg-PSZ, has emerged as an important structural ceramic. Mg-PSZ's can have toughnesses up to 15 MPam$^{1/2}$, the highest known toughness in monolithic ceramics [1]. The existence of R-curve behavior in Mg-PSZ is also well established. R-curve behavior entails increasing crack resistance with crack extension. This effect gives rise to a novel and interesting flaw tolerance; the strengths of these tough Mg-PSZ materials are relatively insensitive to both processing and machining flaws [2,3].

Readey, Steinbrech, and Heuer [4] have recently shown that crack bridging, due to crack surface interaction occurring significantly behind the crack tip, can also lead to additional toughening in Mg-PSZ (such bridging has also been

Mat. Res. Soc. Symp. Proc. Vol. 78. ᶜ1987 Materials Research Society

observed in coarse-grained alumina's [5,6]). Bridging is much more prevalent in the lower toughness Mg-PSZ's (which still have a very high toughness of 9-10 MPam$^{1/2}$), whereas the higher toughness PSZ's (15 MPam$^{1/2}$) show relatively little bridging [7]. The purpose of this paper is to report the differences in R-curve behavior in several PSZ's, and to study how the crack bridging phenomenon varies in materials of different thermal history, and hence different microstructure.

PROCEDURE

The PSZ's used in this study are commercially available 3.0^W/o and 3.4^W/o Mg-PSZ's. All materials were sintered at high temperatures (~1700°C) and given a controlled cooling to develop the metastable, transformable tetragonal precipitate microstructure [8,9]. Two of the 3.4^W/o PSZ's were also given an additional heat-treatment at 1100°C for 2 and 8 hours respectively. The purpose of this lower temperature ("sub-eutectoid") heat-treatment is to enhance the transformability of the tetragonal precipitates [10,11].

Typically, R-curve determinations are measured with specimen geometries which allow for stable crack growth under an applied stress [12], for example, the double-cantilever beam (DCB) geometry. To determine the crack resistance, R, or the toughness, K_R, the crack length, load, and displacement, are measured and incorporated into well-known fracture mechanics solutions for DCB specimens loaded in tension [13,14]. Such an approach was initially used in this work, the load being applied and measured using a standard testing machine[+]. The displacement was determined by mounting a clip gauge directly onto the DCB sample, thereby measuring the crack opening displacement (COD) at the end of the sample. The COD at the point of loading was then calculated from simple trigonometric considerations. The crack length was determined using a traveling microscope attached to a sliding potentiometer. Resolution in crack length measurements was approximately 5μm.

[+]Instron Corporation, Canton, Massachusetts

A starter crack was introduced into the samples using a 500μm diamond saw. An additional 2mm of notch length were produced with a thin, 50μm diamond blade, to attempt to minimize adverse effects due to a too-blunt notch. Samples had dimensions of 25x75x2mm, with a notch depth of 15-20mm.

To minimize possible effects due to the notching process, it was determined that heat-treating the samples at 1000°C for 20 minutes (prior to testing) re-transformed most of the t-ZrO$_2$ precipitates which may have transformed during notching.

A groove approximately 1/3 the thickness of the sample was produced on one side over the length of the sample with a 500μm diamond blade to guide the crack. However, the presence of the groove influenced crack propagation, as was determined by injecting dye penetrant into the crack after some millimeters of crack extension. After the dye dried, the crack was propagated until the sample failed. Inspection of the two DCB halves revealed that the crack was nearly 1-1.5mm longer on the grooved than on the ungrooved side, as shown in Figure 1a. Similar results have also been observed in grooved alumina samples [15].

Steinbrech has shown [15] that grooving both sides of alumina samples results in a straight crack front, but makes accurate crack length measurement impossible. We believe that the deep, single groove introduces a bending component in the thickness direction of the specimen. This subjects the crack tip to a higher tensile stress (hence, larger crack driving force) on the grooved surface than on the ungrooved surface, resulting in an oblique crack front. This is analogous to the oblique crack front typically found in grooved, double torsion specimens [16]. Thus, large and systematic errors in crack length measurement appear inherent in grooved samples, which therefore cannot be used in R-curve determinations.

Some ungrooved samples were tested with the aim of avoiding this problem, and to eliminate possible effects of the t → m transformation during grooving affecting subsequent R-curve determination. During testing of these ungrooved samples, the crack invariably propagated out of the center of the sample, and failure (unstable crack growth) rapidly ensued. Thus, reliable toughness values could not be determined from such ungrooved samples.

To overcome such difficulties, a new sample geometry was employed. As pointed out by Atkins and Mai [17], straight cracks can be obtained in short-DCB samples, obviating the need for crack guiding grooves. Short-DCB samples have a width-length ratio of about unity, with a crack length-sample length ratio of approximately 0.5.

Preliminary tests with this sample geometry (size = 25x30x2mm) resulted in straight crack fronts (Figure 1b), and stable crack growth of several millimeters, enough to reliably determine R-curves. The disadvantages of such a specimen geometry are that the well-known fracture mechanics solutions to the DCB geometry are no longer valid. This obstacle can be circumvented provided that the assumption of linear elastic fracture mechanics (LEFM) is maintained. The fundamental relation for the crack resistance

$$R = \frac{P^2}{2B} \frac{dC}{da} \tag{1}$$

where P is the load, B is the sample thickness, and dC/da is the change in compliance with crack extension, can be used to determine R and K_R. Thus, one needs only to determine dC/da and measure P and B (which is straightforward) to determine R. The critical term, dC/da, can be determined numerically by fitting the compliance (d/P, where d is the displacement) data

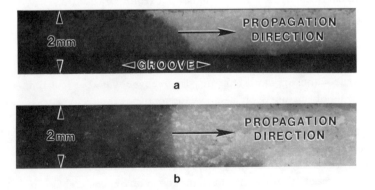

Figure 1. (a)Standard grooved DCB specimen with oblique crack front; (b)"Short" DCB geometry with straight crack front. Dark regions are due to dye penetrant.

to the measured crack length using standard polynomial fitting routines. K_R, is then determined from

$$K = (E'R)^{1/2} \qquad (2)$$

where E' is equal to E for plane stress and $E/(1-\nu^2)$ for plane strain, where ν is Poisson's ratio and E is Young's modulus.

RESULTS

The modified short-DCB testing geometry was successful in determining R-curves of Mg-PSZ. Stable crack growth was maintained over a long enough distance (~5mm) so that the polynomial fitting routine could accurately fit the compliance versus crack growth data. However, the most stable crack growth occurred in materials with the highest plateau toughness. With materials having lower peak toughnesses, the crack would often become unstable as the crack approached the end of the sample, and this data could not be used in the analysis. Thus the utility of the short-DCB testing geometry might be limited to tougher materials.

The load-displacement curves for several specimens are recognizably different, as illustrated in Figure 2. Crack growth for all samples begins before the peak in the P-d curve is attained, typical of materials which show R-curve behavior. For the 1100°C-8hr sample, no sharp maximum was observed as in the lower toughness PSZ's, indicating the material is still toughening even after several millimeters of crack extension.

The R-curves were determined from such P-d curves and measured crack lengths and are shown in Figure 3. Note that in general, the toughness increases rapidly within the first few hundred microns of crack extension and gradually levels off to a plateau toughness. Initial K_R's are on the order of 5-7 MPam$^{1/2}$. The 3.0^W/o material has a plateau value of approximately 9 MPam$^{1/2}$, whereas the 3.4^W/o materials are tougher, with plateau values between 10 and 18 MPam$^{1/2}$.

Some samples of each type were also mildly etched (90 seconds in 30% HF) (Figure 4) prior to testing to reveal the grain boundary regions and the mode of crack propagation

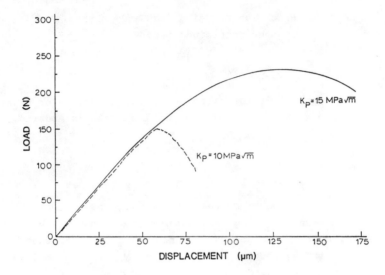

Figure 2. Typical load-displacement curves for PSZ's with
two different plateau toughnesses (K_P).

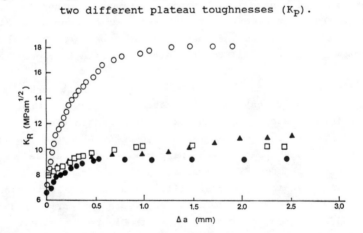

Figure 3. K_R-curves versus crack extension. The toughness
increases rapidly in the first several hundred
microns of crack extension, and then levels off
to a plateau value.

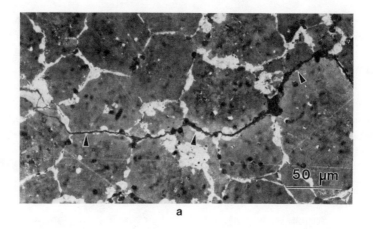

a

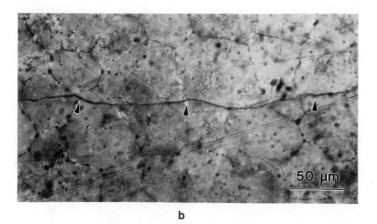

b

Figure 4. (a) Low plateau toughness PSZ illustrating
significant intergranular fracture. (The highly
reflective regions reveal the grain boundaries.)
(b) Transgranular crack propagation in the higher
toughness PSZ's. The crack paths are arrowed.

(inter- or transgranular). A significant degree of
intergranular fracture occurred in the low toughness 3^W/o-PSZ.
In the intermediate toughness materials, less intergranular
failure occurred, while in the highest toughness material, the
crack path was almost entirely transgranular.

Crack propagation was also unusual in the low toughness
PSZ's. Cracks often arrested, with a secondary microcrack
nucleating at another grain boundary, often several grains
away. The secondary crack would begin to propagate, with the
material left behind linking or "bridging" the two halves of
the crack surfaces (Figure 5). The bridge would eventually
fail after another 1-2 mm of crack extension.

The highest toughness PSZ did not show the same crack
propagation behavior as the lower toughness PSZ's. Rather,
the crack propagated in a slow, continuous fashion, without
arresting and without subsequent bridge formation.

Optical microscopy using Nomarski interference was
employed to observe the crack tip transformation zone
(Figure 6). The plateau toughness can be correlated with the
zone thickness -- a zone thickness of approximately 10-25μm
corresponded to a peak toughness of 9 MPam$^{1/2}$, while a zone
thickness of several hundred microns was present for the
toughest (18 MPam$^{1/2}$) material.

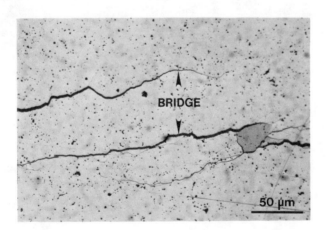

Figure 5. Optical micrograph illustrating bridge formation
behind the crack tip.

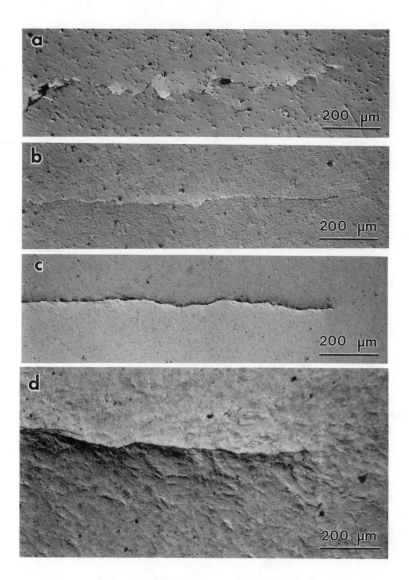

Figure 6. Optical micrographs (Nomarski interference
contrast) revealing transformation zone widths.
The lowest plateau toughness PSZ is shown in (a);
peak toughness increases from (b) to (c) to (d),
and correlates with transformation zones of 10μm
in (a) to 200μm in (d).

DISCUSSION

One interesting result from these tests is the strong
correlation between plateau toughness, transformation zone
size, and mode of crack propagation (inter or transgranular).
The highest K_R PSZ exhibited primarily transgranular fracture,
whereas the PSZ with the lowest K_R shows a significant amount
of intergranular fracture. The highest toughness PSZ had the
largest transformation zone size, nearly 200 microns; the PSZ
with the lowest toughness had an estimated zone size of only
10-20 microns.

All theories of transformation toughening indicate that
the plateau toughness scales with the square root of the zone
size [18]. From strictly zone size considerations, the plateau
K_R results are consistent with the transformation zone sizes
estimated from the Nomarski micrographs.

Bridging is most likely to occur when a significant
fraction of the crack path is intergranular (Figures 4 and 5).
Bridging thus appears directly related to the mode of crack
propagation. However, what determines the mode of fracture is
still not fully understood. Bridges typically nucleate at
grain boundaries, indicating that grain boundaries in the less
tough materials have a lower toughness than the bulk. Grain
boundary impurity phases containing SiO_2, such as a silicate
glass and forsterite, Mg_2SiO_4, have been observed in these
PSZ's with transmission electron microscopy (TEM) [19]. The
presence of silica at grain boundaries is known to increase
the grain boundary monoclinic content, both by leaching MgO
from surrounding grains [20], and by enhancing the eutectoid
decomposition reaction [21,22]. Swain and Hannink [23] have
demonstrated that the presence of a grain boundary monoclinic
ZrO_2 phase causes intergranular fracture. The SiO_2 content in
the lower toughness materials (exhibiting more bridging) is
known from TEM to be greater than in the tougher PSZ's. Thus,
bridging and intergranular fracture appear to be related to
the weakening effect of grain boundary monoclinic ZrO_2 and
impurity phases. However, the two 3.4^W/o PSZ's, with
identical chemistries but different heat treatments, underwent
different degrees of transgranular fracture, varying inversely
with the plateau toughness.

The occurrence of bridging in PSZ's with the lower toughnesses is contrary to that observed in alumina ceramics, where R-curve behavior arises entirely from the bridging effect [5,6]. The disparate nature of bridging in alumina and PSZ ceramics can be rationalized as follows: In alumina, as in PSZ, bridging results from extensive intergranular fracture and leads to interlocking grains which actually bridge -- linking the crack surfaces together behind the crack tip. This bridging provides a restraining force which reduces the crack tip stress intensity and hence increases the toughness [5,6].

However, in PSZ's, it is the interaction between bridging and the transformation which may be important. For example, consider a crack propagating with a transformation zone size on the order of 10 microns. As the main crack propagates, ascending the K_R-curve, a secondary crack which nucleates at a grain boundary 1-2 grains away (grain size ~75μm) lies completely outside the transformation zone of the main crack. If the secondary crack now begins to propagate, it develops with its own K_R-curve, independent of the history of the previous crack; the K_R-curve re-initiates with each secondary crack. (Experimentally, the crack length was taken as the total crack length, including secondary cracks. The overall result is that the K_R-curve represents the average curve). Therefore the maximum plateau toughness due to transformation toughening may never be attained while the secondary cracks nucleate at a lower K_R value.

The absence of bridging in the toughest PSZ's can also be explained using a similar argument. In these materials, the transformation zone is significantly larger than the grain size. Therefore, the most probable regions of secondary crack nucleation are sufficiently far from the main crack tip that formation of these cracks is unlikely. Due to the dilatational component of the transformation, the region contained within the transformation zone is in compression. Thus, formation of cracks within the zone is inhibited by the compressive forces from the transformation zone itself. The main crack then propagates with a single K_R-curve which experiences the full effect of transformation toughening.

This also explains the uniformity of the K_R-curve in the toughest PSZ's.

This suggests that intergranular fracture and subsequent bridging effect are strongly influenced by transformation zone size. As the zone size decreases, the amount of intergranular fracture increases, and the effects of bridging become more important as a toughening mechanism. Conversely, a large transformation zone results in nearly transgranular crack propagation, leading to transformation toughening as the primary toughening mechanism. The increase in toughening due to bridging is small compared to transformation-toughening -- on the order of 2-4 MPam$^{1/2}$ in alumina. The increase is probably less in PSZ's, since fracture is never entirely intergranular as in the alumina's which show K_R behavior. Transformation toughening is a much more potent mechanism of toughening, with increases in toughness of 5-12 MPam$^{1/2}$ being common.

Bridging and transformation appear to be opposing mechanisms in PSZ, suggesting that eliminating bridging through transgranular crack propagation will result in higher toughness PSZ's. The almost entirely transgranular behavior of the toughest PSZ suggests that a large transformation zone has the effect of preventing secondary cracks, or bridges, from developing and propagating. This may be accomplished by the proper heat-treatment during processing. In fact, if zone size can be adequately controlled, it may be possible to compensate for deleterious grain boundary phases by having zone sizes considerably larger than the grain size, thereby "homogenizing" the microstructure with respect to grain boundary structure.

One other aspect of large transformation zones deserves mention at this point, as it involves the assumption of linear elasticity, which was assumed throughout this paper. Such an assumption allows use of the fundamental equation (1), and compliance vs. crack length fits in determination of R and K_R. However, as demonstrated by Burns and Swain [1] and Marshall [2], the highest K_R PSZ's, which show extensive transformation zone sizes (several hundred microns, a significant fraction of the sample thickness during the test), also exhibited considerable non-linearity in the stress-strain behavior. During our experiments, unusually large crack opening displacements (COD's) were necessary for crack propagation. Such large transformation zones and large COD values are indicative that extensive plastic deformation is occurring near the crack tip. This effect is almost certainly due to "transformation plasticity", i.e. the special type of crystal plasticity associated with the martensitic transformation in ZrO_2. In materials with significant plasticity, equation (1) becomes an insufficient criterion for determining R. In fact, R and K_R are no longer valid material parameters, and J concepts (elastic-plastic fracture criteria) must be employed to accurately describe the crack resistance. Unfortunately, measurement techniques of crack resistance for PSZ's exhibiting extensive transformation plasticity, and which adequately take into account bridging, have yet to be developed. This area, as well as the relationship between bridging and transformation toughening, are currently under investigation.

CONCLUSIONS

The fracture resistance of Mg-PSZ's were investigated using a modified, short-DCB testing geometry. The mode of crack propagation is important in these materials in determining whether the full effect of transformation toughening is realized. Materials which undergo primarily transgranular behavior exhibit the maximum toughness, while materials with lower plateau toughness exhibit intergranular fracture. Simultaneously, and perhaps more importantly, the toughest materials also exhibit the largest transformation zone sizes. We suggest the larger transformation zone prevents secondary crack propagation (bridging) from occurring.

120

ACKNOWLEDGEMENTS

We thank Dr W. Schaarwächter at the University of
Dortmund for many useful discussions and for extensive use of
the facilities at his disposal. M.J. Readey would also like
to thank Andreas Reichl and Frederike Deuerler for their help
and insight into R-curve measurements in ceramics. Financial
support from NSF grant numbers DMR 82-14128 and INT-8610359 is
greatly appreciated. M.J. Readey is the TIMKEN Fellow at Case
Western Reserve University.

REFERENCES

1. S.J. Burns, M.V. Swain, J. Am. Ceram. Soc. <u>69</u> 22 (1986).
2. D.B. Marshall, J. Am. Ceram. Soc. <u>69</u> 173 (1986).
3. M.J. Readey, A.H. Heuer, R.W. Steinbrech, unpublished
    work.
4. M.J. Readey, R.W. Steinbrech, A.H. Heuer, "Influence of
    Microstructure on R-Curve Behavior in Mg-PSZ", presented
    at the 88th Annual Meeting of the American Ceramic
    Society, Chicago, Il April 30, 1986.
5. R. Knehans, R.W. Steinbrech, J. Mat. Sci. Letters <u>1</u> 327
    (1982).
6. R.F. Cook, B.R. Lawn, and C.J. Fairbanks, J. Am. Ceram.
    Soc. <u>68</u> 604 (1985).
7. A.H. Heuer, submitted, J. Am. Ceram. Soc.
8. D.L. Porter, A.H. Heuer, J. Am. Ceram. Soc. <u>62</u> 298 (1979).
9. R.H.J. Hannink and M.V. Swain, J. Aust. Ceram. Soc
    <u>18</u> 53 (1982).
10. R.H.J. Hannink, and R.C. Garvie, J. Mater. Sci. 17 2637
    (1982).
11. R.H.J. Hannink, J. Mater. Sci. <u>18</u> 457 (1983).
12. D. Broek, <u>Elementary Fracture Mechanics</u>, 3rd ed.
    (Martinus Nijhoff Publishers, Boston, 1982).
13. S.M. Wiederhorn, in Fracture Mechanics of Ceramics,
    Vol. 2, edited by R.C. Bradt, D.P.H. Hasselman and F.F.
    Lange, Plenum Press, New York, 1974.
14. N. Bathena, R.G. Hoagland, and G. Meyrick, J. Am.
    Ceram. Soc. <u>67</u> 799 (1984).
15. R.W. Steinbrech and F. Deuerler, to be published.
16. G.G. Trantina, J. Am. Ceram. Soc. <u>60</u> 338 (1977).
17. A.G. Atkins and Y.W. Mai, <u>Elastic and Plastic Fracture</u>,
    (John Wiley & Sons, New York, 1985), pp. 198-204.
18. R. McMeeking, A.G. Evans, J. Am Ceram. Soc. <u>65</u> 242 (1982).
19. M.J. Readey, A.H. Heuer, unpublished work.
20. R. Chaim and D.G. Brandon, J. Mater. Sci. <u>19</u> 2934 (1984).
21. J. Drennan, R.H.J. Hannink, J. Am. Ceram. Soc. <u>69</u> 541
    (1986).
22. S. Farmer, A.H. Heuer, and R,H,J, Hannink, in press.
23. M.V. Swain and R.H.J. Hannink, Advances in Ceramics,
    Vol. 12, edited by N. Claussen, M. Rühle, and A.H.Heuer,
    American Ceramic Society, Columbus, 1984.

PART III

Mechanical Properties and Microstructures of Zirconia Toughened Ceramics

MECHANICAL PROPERTY AND MICROSTRUCTURE OF
TZP AND TZP/Al$_2$O$_3$ COMPOSITES

KOJI TSUKUMA AND TSUTOMU TAKAHATA
Tokyo Research Center, Toyo Soda Manufacturing, Company, Hayakawa,
Ayase-shi, Kanagawa 252, Japan

ABSTRACT

There are various types of microstructure in the TZP materials and
their composite materials. The fine-grained microstructure is well-known
as the basic microstructural type of TZP materials. Y-TZP and Ce-TZP belong
to this group. The large-grained microstructure can exist in the TZP consisting of the tetragonal phase with low metastability, that is, Y-TZP doped
with TiO$_2$ and Ce-TZP containing a high CeO$_2$ content. The composite material
between Y-TZP and β-lanthanum alumina possessed a unique microstructure
including elongated grains of β-Al$_2$O$_3$ type structure. This study provides a
summary of the mechanical properties of these TZP and β-Al$_2$O$_3$ type structure
composite materials, and points out how the mechanical behavior depends on
the microstructural features.

INTRODUCTION

Y$_2$O$_3$-stabilized tetragonal zirconia polycrystal (Y-TZP) ceramic is
a toughened zirconia consisting of fine grained microstructure. Gupta
et al., [1] first investigated this material. They indicated that the
tetragonal phase can be retained by decreasing its grain-size, and the
resultant material exhibits high fracture toughness. Lange [2] revealed
that fracture toughness of Y-TZP increases with decreasing Y$_2$O$_3$ content,
and his experimental result is consistent with the concept of stress-
induced transformation toughening. The mechanical properties of Y-TZP
have been studied by many investigators [3-6], as a result, it was
recognized that this material is a high strength ceramic. The high
strength may be due to the contribution of both transformation toughening
and the homogeneous fine-grained microstructure.
    Recently, it was reported that the strength of Y-TZP is enhanced by
the addition of Al$_2$O$_3$ [7]. This composite material, Y-TZP/Al$_2$O$_3$, exhibited
an extremely high strength of >2 GPa.
    CeO$_2$-stabilized tetragonal zirconia polycrystals (Ce-TZP) are an
attractive material from a view-point of transformation toughening. The
CeO$_2$ content in tetragonal phase can be varied widely, which forms a
striking contrast to the narrow range of Y$_2$O$_3$ content in Y-TZP. Recently,
the mechanical properties of Ce-TZP has been reported, and special attention
was given to the high fracture toughness and the transformation behavior of
this material [8].
    In this study, the mechanical properties of pure and composite type TZP
materials are summarized, and the mechanical behavior is discussed based on
the microstructural aspects.

MATERIALS

The TZP materials (Ce-TZP and Y-TZP) and TZP/Al$_2$O$_3$ composite
materials (Ce-TZP/Al$_2$O$_3$ and Y-TZP/Al$_2$O$_3$) were fabricated by the sintering or
isostatically hot-pressing. The sintering was performed at low temperatures
(1400–1600°C), and the isostatically hot-pressing was conducted at 1500°C

Mat. Res. Soc. Symp. Proc. Vol. 78. ⓒ 1987 Materials Research Society

and 100 MPa in argon gas. Dense samples of Ce-TZP and Ce-TZP/Al$_2$O$_3$ could not be obtained by isostatically hot-pressing owing to the destabilization resulting from the reduction of CeO$_2$ in argon atmosphere.

A composite material consisting of Y-TZP and La$_2$O$_3$.11Al$_2$O$_3$ was obtained by using a mixed powder of ZrO$_2$ (2 mol % Y$_2$O$_3$), Al$_2$O$_3$ and La$_2$O$_3$. The β-alumina type compound, La$_2$O$_3$.11Al$_2$O$_3$ (β-La-Al$_2$O$_3$), was formed at a sintering temperature of 1450°C by following reaction.

$$10 \ \alpha\text{-Al}_2\text{O}_3 + \text{La}_2\text{O}_3.\text{Al}_2\text{O}_3 \rightarrow \beta\text{-La}_2\text{O}_3.11\text{Al}_2\text{O}_3 \ \dots \tag{1}$$

The Ce-TZP containing 16 mol % CeO$_2$ and the Y-TZP doped with 15 mol % TiO$_2$ were sintered at a temperature of 1700°C to promote a grain-growth. The characteristics of these various types of TZP based materials are shown in Table 1 and 2.

Table 1. Characteristics of TZP materials and TZP/Al$_2$O$_3$ composites

Y$_2$O$_3$	CeO$_2$	Al$_2$O$_3$ (wt %)	Phase content	Grain-size (μm)	Fabrication
(mol %)					
			Y-TZP		
2.0	—	—	t	0.2 & 0.5	Sintered
2.5	—	—	t + c	"	at 1400°C
3.0	—	—	"	"	and 1500°C
3.5	—	—	"	"	HIP'ed at
4.0	—	—	"	"	1500°C
			Y-TZP/Al$_2$O$_3$		
2.0	—	10	t + α-Al$_2$O$_3$	0.5	Sintered
"	—	20	"	"	at 1500°C
"	—	40	"	"	HIP'ed at
"	—	60	"	"	1500°C
"	—	80	"	"	
			Ce-TZP		
—	12.0	—	t	0.5,1.0,2.5	Sintered
—	14.0	—	"	"	at 1400,
—	16.0	—	"	"	1500 &
					1600°C
			Ce-TZP/Al$_2$O$_3$		
—	12.0	10	t + α-Al$_2$O$_3$	1.0 & 2.5	Sintered
—	"	20	"	"	at 1500°C
	"	40	"	"	& 1600°C
	"	60	"	"	
	"	80	"	"	

Table 2. Characteristics of the large grained TZP
and Y-TZP/β-La-Al$_2$O$_3$ composites

Y$_2$O$_3$	CeO$_2$	TiO$_2$	β-La-Al$_2$O$_3$ (wt %)	Phase content	Grain-size (μm)	Fabrication
(mol %)						
				Ce-TZP		
—	16	—	—	t	10	Sintered at 1700°C
				Y-TZP + TiO$_2$		
2	—	15	—	t	20	"
				Y-TZP/β-Al$_2$O$_3$		
2	—	—	40	t + β-La-Al$_2$O$_3$	0.5	Sintered and HIP'ed at 1500°C

MICROSTRUCTURAL FEATURES

The microstructures of various types of TZP-based materials are
schematically illustrated in Fig. 1. It is recognized that the tetragonal
phase can not be retained when the grain-size exceeds the critical value.
Therefore, TZP materials are usually fabricated at low sintering temperatures, below 1600°C, to avoid a grain-coarsening. As a result, TZP
materials usually possess a fine-grained microstructure, type A-1 or A-2
(Fig. 1).

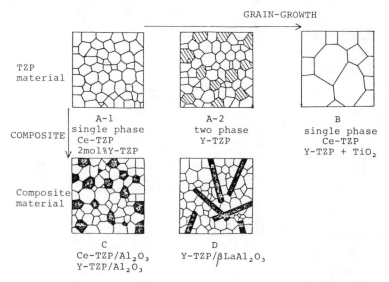

Fig. 1. Various types of microstructure of TZP and TZP
composite materials.

However, it is possible to increase the critical grain-size by controlling the stabilizer components in the tetragonal phase. As shown in previous reports [8-9], the addition of Ce_2O or TiO_2 is effective for stabilizing the tetragonal phase. Thus, TZP materials with large-grained microstructure of type B can be obtained in the system of ZrO_2-Y_2O_3-TiO_2 or ZrO_2-CeO_2.

In the composites consisting of TZP plus other materials, there is the possibility of changing the morphology of second phase materials. The microstructure of type D includes elongated grains, Fig. 1. This microstructure is made by the in situ method utilizing the anisotropic grain-growth of second phase material. It is known that β-alumina is a material which readily exhibits such anisotropic grain-growth. Thus, the composites with the microstructure of type D can be obtained in the system between TZP and the β-alumina type compound.

Fine-Grained Microstructure

It is well-known that the Y-TZP materials possesses a fine-grained microstructure. As shown in the previous report [2], the critical grain-size of Y-TZP containing 2–3 mol % Y_2O_3 is below 1 μm. Y-TZP containing >2 to 5 mol % is composed of a mixture of two phases (cubic + tetragonal), while the 2 mol % Y_2O_3-TZP consists of a single tetragonal phase.

The Ce-TZP materials belong to the same microstructure type as Y-TZP. The microstructure of Ce-TZP is shown in Fig. 2(a). The materials obtained by a conventional sintering at temperatures of 1400–1600°C show a grain-size of 0.5–2.5 μm, which corresponds an upper limit which is about twice as large as that of Y-TZP. The Ce-TZP is a single tetragonal phase material with a wide range of CeO_2 content, between 12 and 20 mol %. The previous study [10] indicated that the critical grain-size of 12 mol % CeO_2-TZP is about 3 μm.

The composites between TZP and α-Al_2O_3 belong to the fine-grained microstructure group. The microstructure of Ce-TZP/Al_2O_3 is shown in Fig. 2(b). The addition of Al_2O_3 is effective for reducing a grain-growth as shown in Fig. 2(a) and (b). Similar phenomena was observed in the two phase cubic + tetragonal Y-TZP [11].

Large-Grained Microstructure

Higher sintering temperature can be used to densify Ce-TZP containing more than 16 mol % CeO_2 because of its large critical grain-size. It was estimated that the critical grain-size of 16 mol % Ceo_2-TZP is more than 10 μm. The microstructure of this material sintered at 1700°C is shown in Fig. 2(c). Also, the 2 mol % Y-TZP doped with 15–20 mol % TiO_2 has a large critical grain-size. The TiO_2 doping is effective in promoting grain-growth, and also in stabilizing the tetragonal phase [12]. These TZP materials consist of a single tetragonal phase. It is well-known that the two phase, cubic + tetragonal, material results in the PSZ type microstructure containing fine tetragonal precipitates by high temperature sintering [13].

Microstructure Including the Elongated Grains

The Y-TZP/β-$LaAl_2O_3$ composites have a unique microstructure. The microstructure of the composite, 2 mol % Y-TZP + 40 wt % β-$LaAl_2O_3$, is shown in Fig. 3. The β-$La_2O_3 \cdot 11Al_2O_3$ particles form elongated grains through the reaction of Eq. (1) above 1450°C. As shown in Fig. 4, this particle was a single crystal with included zirconia fine particles, and the cleavage plane (001) was parallel with a long axis.

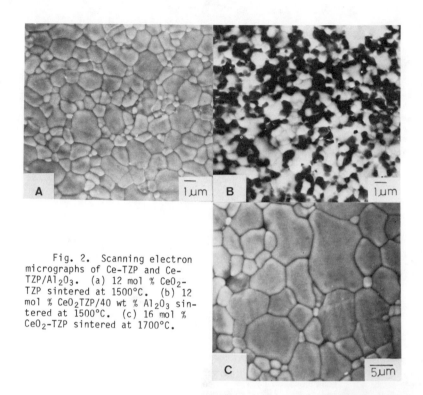

Fig. 2. Scanning electron micrographs of Ce-TZP and Ce-TZP/Al$_2$O$_3$. (a) 12 mol % CeO$_2$-TZP sintered at 1500°C. (b) 12 mol % CeO$_2$TZP/40 wt % Al$_2$O$_3$ sintered at 1500°C. (c) 16 mol % CeO$_2$-TZP sintered at 1700°C.

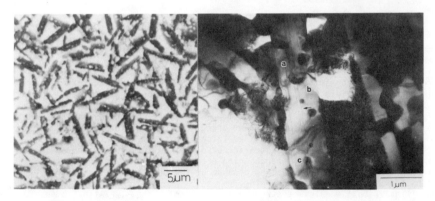

Fig. 3. Scanning electron micrographs of the composite, 2 mol % Y$_2$O$_3$-TZP/40 wt % La-β-Al$_2$O$_3$, sintered at 1500°C.

Fig. 4. Transmission electron micrograph of composite 2 mol % Y$_2$O$_3$-TZP/40 wt % La-β-Al$_2$O$_3$.

STRENGTH AND TOUGHNESS

Fine-Grained Microstructure

Previous studies [3,7] reported on the bending strength and fracture toughness of the fine-grained TZP materials and TZP/Al_2O_3 composite materials. The results are summarized in Figs. 5—7. As shown in Fig. 5, a material with very high toughness was found in the Ce-TZP group. The previous work [8] indicated that the transformability of tetragonal phase increases with decreasing CeO_2 content and with increasing grain-size, which corresponds to an increase in fracture toughness. The high fracture toughness of the 12 mole % CeO_2-TZP resulted from such high transformability. This suggests that the tetragonal phase in 12 mol % CeO_2-TZP can be retained in a very metastable state.

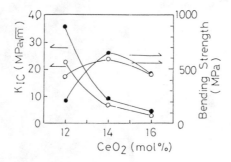

Fig. 5. Bending strength and fracture toughness of Ce-TZP sintered at 1400°C (o) and 1500°C (•).

As shown in Fig. 7, material with extremely high strength was found in a series of Y-TZP/Al_2O_3 composites. The Y-TZP/20 wt % Al_2O_3 composite material exhibited a strength of 2.4 GPa. The previous work [14] reported that the defects in this material are very small (about 10 μm) in the isostatically hot-pressing composite. It is obvious that the high strength resulted from such small defect-size. Swain [15] also proposed that increasing the yield strength is a necessary condition to attain high strength. This corresponds with an increase in the critical stress required for the martensitic phase transformation. This critical stress is enhanced by the Al_2O_3 addition because of the strong constraining force of the matrix containing Al_2O_3. Thus, the Al_2O_3 addition effectively enhanced the strength of the Y-TZP/Al_2O_3 system and also the Ce-TZP/Al_2O_3 system as shown in Fig. 7. However, this strengthening effect reduces when Al_2O_3 content increases beyond a certain value, because of the decrease in the contribution from transformation toughening.

The Weibull modulus of strength in the 3 mol % Y_2O_3-TZP/Al_2O_3 composite is shown in Fig. 8. The two materials, A and B, were prepared from the different batches of powders. The material A showed a low Weibull modulus owing to the presence of a low strength region, but the sample B showed high modulus, 14. The fracture origins observed on the weak test bars of the material A are shown in Fig. 9. Agglomerates of Al_2O_3 particles or flaws in the surface was responsible for low strengths. This suggests that very precise control of processing defects is required to maintain the reliability of strength. It appears that the rather low fracture toughness of this material intensifies the flaw sensitivity.

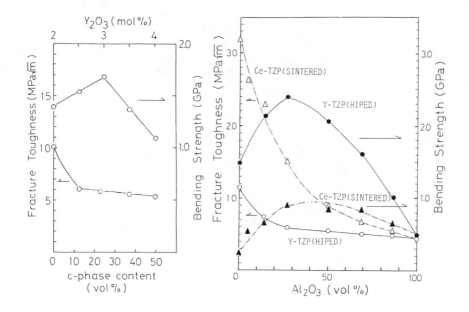

Fig. 6. Bending strength and fracture toughness of Y-TZP isostatically hot-pressed at 1500°C and 100 MPa.

Fig. 7. Bending strength and fracture toughness of Y-TZP/Al₂O₃ isostatically hot-pressed at 1500°C and 100 MPa

The 12 mol % CeO₂-TZP showed low average strength, 380 MPa, but the Weibull modulus of strength was high, 40. The previous work [10] indicated that this material exhibited significant plastic deformation prior to failure under low applied stress, which was attributed to the large transformed zone. The fracture strength is governed by this yielding behavior. The high fracture toughness of this material is also helpful in obtaining highly reliable strength.

The relation between bending strength and fracture toughness obtained for the fine-grained TZP and TZP/Al₂O₃ is shown in Fig. 10. All groups of TZP and TZP/Al₂O₃ exhibited a maximum in strength at intermediate values of fracture toughness. Swain [15] describe such strength/toughness behavior in transformation toughened zirconia by assuming a small scale yielding model. The results of Fig. 10 indicate that this concept of strengthening versus toughening is consistent with this model and is similar to the behavior of metallic materials.

Large-Grained Microstructure

The bending strength and fracture toughness of the large-grained TZP materials are listed in Table 3. The strength of these materials was not as high as that of the fine-grained TZP materials. However, the fracture toughness was increased considerably by the grain coarsening. The increment of increased fracture toughness results from the increase in transformability of tetragonal phase due to grain-growth.

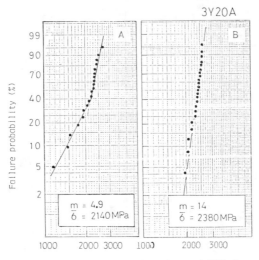

Fig. 8. Weibull probability of strength of the composite 3 mol % Y_2O_3-TZP/20 wt % Al_2O_3. (a) Wide distribution of strength. (b) narrow distribution of strength.

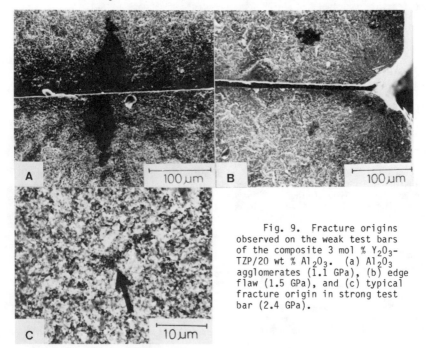

Fig. 9. Fracture origins observed on the weak test bars of the composite 3 mol % Y_2O_3-TZP/20 wt % Al_2O_3. (a) Al_2O_3 agglomerates (1.1 GPa), (b) edge flaw (1.5 GPa), and (c) typical fracture origin in strong test bar (2.4 GPa).

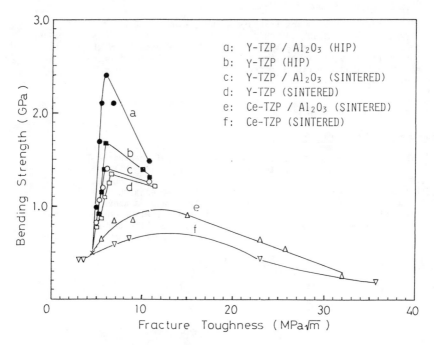

Fig. 10. Relationship between strength and toughness obtained for the TZP and TZP/Al$_2$O$_3$ composites.

Table 3. Strength and toughness of the large-grained TZP and TZP/β-Al$_2$O$_3$.

	Bending strength (MPa)	Fracture toughness (MPa $\sqrt{m}$)
Ce-TZP	470	12.3
2 mol % Y$_2$O$_3$-TZP + 15 mol % TiO$_2$	560	15.0
2 mol % Y$_2$O$_3$-TZP + 40 wt % β-La-Al$_2$O$_3$	780 (Sintered)	7.8
	1300 (HIP'ed)	

Microstructure Including the Elongated Grains

The bending strength and fracture toughness of the 2 mol % Y_2O_3-TZP/40 wt % β-LaAl$_2O_3$ composite are shown in Table 3. It is generally recognized that the β-alumina type compound is weak because of the cleavage fracture related to the layer structure. However, the addition of β-LaAl$_2O_3$ did not result in a remarkable decrease in strength as compared to TZP and TZP/α-Al$_2O_3$ composites.

HIGH TEMPERATURE STRENGTH AND DEFORMATION

Fine-Grained Microstructure

The temperature dependences of strength of the fine-grained TZP materials and TZP/Al$_2O_3$ composites are shown in Fig. 11. Y-TZP and Ce-TZP showed low strength at high temperature, 300 MPa at 800°C. The isostatically hot-pressed Y-TZP/Al$_2O_3$ composites showed high strength at high temperature 700—1000 MPa at 1000°C. At temperatures above 1000°C, the strength of Y-TZP/Al$_2O_3$ increased with increasing Al$_2O_3$ content. The Al$_2O_3$ addition is effective in increasing strength at high temperature.

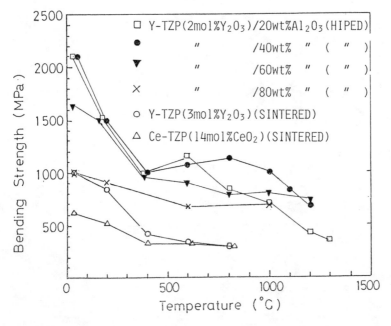

Fig. 11. Temperature dependence of strength of TZP materials and TZP/Al$_2O_3$ composite materials.

The stress/strain curves of Y-TZP and Y-TZP/Al$_2O_3$ from high temperature bending test are shown in Fig. 12. The 3 mol % Y_2O_3-TZP showed large plastic deformation at 1200°C under a slow loading rate. Wakai et al. [16] reported that the fine-grained TZP materials exhibited super plasticity at high temperature. The Al$_2O_3$ addition decreased the amount of plastic deformation.

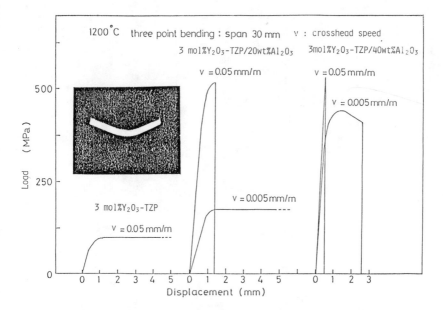

Fig. 12. Stress/strain curve of Y-TZP and Y-TZP/Al₂O₃ in high temperature bending test. (All samples were isostatically hot-pressed at 1500°C and 100 MPa.)

Microstructure Including the Elongated Grains

The sintered Y-TZP/β-LaAl₂O₃ composite showed high strength at high temperature, 600 MPa at 1000°C and 500 MPa at 1300°C. There is no rapid decrease in strength at high temperature. The deformation behavior was comparable with that of the fine grained Y-TZP/Al₂O₃ composites in the high temperature bending test. These results are shown in Fig. 13. The volume fraction of β-LaAl₂O₃ is approximately equal to that of α-Al₂O₃ in the TZP/α-Al₂O₃ composites because of the similar densities of α and β alumina. The Y-TZP/β-LaAl₂O₃ composite was not as deformable as the Y-TZP/Al₂O₃ composite. This result suggests that the elongated β-alumina particles are effective in inhibiting plastic deformation.

Large-Grained Microstructure

The stress/strain curves from the high temperature bending tests are compared for 16 mol % CeO₂-TZP with grain-size of 10 μm and 12 mol % CeO₂-TZP with grain-size of 1 μm, Fig. 14. The large-grained Ce-TZP showed no plastic deformation. This indicates that the grain coarsening diminishes plastic deformation at high temperature.

CONCLUSION

The Y-TZP/20 wt % Al₂O₃ composite had extremely high strength and the 12 mole % CeO₂-TZP exhibited high fracture toughness. This indicates that a composite containing a high elastic modulus material which is composed of fine grains is preferable for high strength, and

134

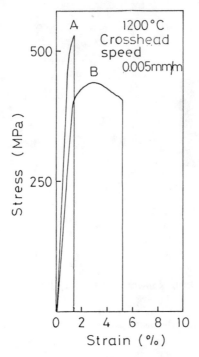

Fig. 13. Stress/strain curve of the
composite 2 mol % Y_2O_3-TZP/40 wt % β-La-Al$_2O_3$
in high temperature bending test. (a) 2 mol %
Y_2O_3-TZP/40 wt % β-La-Al$_2O_3$ sintered at 1500°C.
(b) 2 mol % Y_2O_3-TZP/40 wt % α-Al$_2O_3$ sintered
at 1500°C.

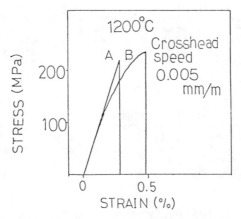

Fig. 14. Stress/strain curves in a bending
test at 1200°C. (a) 16 mol % CeO_2-TZP sintered at
1700°C. (b) 12 mol % CeO_2-TZP sintered at 1450°C.

that the single tetragonal phase material with high metastability is
advantageous for high fracture toughness. Two types of microstructure,
the large-grained microstructure and the microstructure including
elongated grains, were effective for improving the mechanical properties
at high temperature, especially the resistance to high temperature
deformation.

REFERENCES

1. T. K. Gupta, J. H. Betchtold, R. C. Kuzniki, L. H. Kadoff, and
   B. R. Rossing, J. Mat. Sci., 12, 2421-6 (1977).
2. F. F. Lange, J. Mat. Sci., 17, 240-6 (1982).
3. K. Tsukuma and M. Shimada, Am. Soc. Bull., 64(2), 310-13 (1984).
4. M. Matsui, T. Soma, and I. Oda, Advances in Ceramics, Vol. 12,
   Science and Technology of Zirconia II, edited by N. Claussen,
   M. Rühle, and A. H. Heuer, (Am. Ceram. Soc. Inc., 1984), 371–8.
5. M. Watanabe, S. Iio, and I. Fukuura, Advances in Ceramics, Vol 12,
   Science and Technology of Zirconia II, edited by N. Claussen,
   M. Rühle, and A. H. Heuer, (Am. Ceram. Soc. Inc., 1984), 391–8.
6. T. Masaki, J. Am. Ceram. Soc., 69(7), 519-22 (1986).
7. K. Tsukuma, K. Ueda, and M. Shimada, J. Am. Ceram. Soc., 68(1), c-4–5
   (1985).
8. K. Tsukuma and M. Shimada, J. Mat. Sci., 20, 1178–84 (1985).
9. K. Tsukuma, Zirconia Ceramics 8, edited by S. Somiya and M. Yoshimura,
   (Uchida-roho-kaku, Tokyo, 1986) 11–20.
10. K. Tsukuma, Am. Ceram. Soc. Bull., 65(10), to be published (1986).
11. F. F. Lange, J. Am. Ceram. Soc., 69(3) 240–42 (1986).
12. K. Tsukuma, J. Mat. Sci. Lett., to be published (1986).
13. R. H. J. Hannink, J. Mat. Sci., 13, 2487–96 (1978).
14. K. Tsukuma, K. Ueda, and M. Shimada, J. Am. Ceram. Soc., 68(2),
    c-56–58 (1985).
15. M. V. Swain and L. R. F. Rose, J. Am. Ceram. Soc., 69(7), 511–18
    (1986).
16. F. Wakai, S. Sakaguchi, and Y. Matsuno, Adv. Ceram. Mater., 1, 3
    (1986).

EVALUATION OF COMMERCIALLY AVAILABLE TRANSFORMATION
TOUGHENED ZIRCONIA

JEFFREY J. SWAB[*]
* Materials Technology Laboratory, SLCMT-MCC, Watertown, MA 02172

ABSTRACT

Transformation toughened zirconia (TTZ) is a material being
considered for use in advanced heat engines. However, at elevated
temperatures TTZ materials undergo a phase transformation from tetragonal
to the monoclinic with an associated volume increase of approximately 5%.
This transformation results in a loss of strength and fracture toughness.
Six commercially available Japanese TTZ materials and one experimental
domestic grade were examined for the extent and effect of this phase
transformation after exposure to elevated temperatures (1000 to 1200°C)
for times of 100 and 500 hours. Tests completed to date show that all the
TTZ materials examined transform and lose strength, but to various
degrees. Strength losses after heat treatment at 1000°C for 100 and 500
hours, ranged from a high of 60% to as little as 7%. Additional heat
treatments of 500 hours at 1100 and 1200°C were carried out on TTZ's which
had strength losses of 15% or less after exposure to 1000°C.

Although the major thrust of this program is to examine the effects
of high temperatures on TTZ materials, a preliminary examination of the
effects of low temperatures on the properties is also being done. A small
number of specimens are undergoing treatments at 200°C and ~.8 MPa water
vapor pressure for 50 hours. Early indications are that strength is
greatly reduced after this treatment.

INTRODUCTION

A wide variety of transformation toughened zirconias (TTZ) are
currently being examined for structural applications (i.e., engines and
cutting tools) because of their unusual combination of high strength and
fracture toughness. Studies have been conducted on magnesia-partially
stabilized zirconia (Mg-PSZ), yttria tetragonal zirconia polycrystals
(Y-TZP) [1-6], and recently ceria tetragonal zirconia polycrystals
(Ce-TZP) [7]. Their unusual properties stem from a stress-induced,
martensitic tetragonal (t) to monoclinic (m) phase transformation, [8]
hence the name "transformation toughened".

The mechanisms of this toughening are believed to include one or more
of the following:

1) deflection of the crack tip;

2) microcracking which leads to crack branching and an increase
in the energy required for continued crack growth;

3) the absorption of energy by the phase transformation
process. [1]

Mat. Res. Soc. Symp. Proc. Vol. 78. ⁱ 1987 Materials Research Society

Although TTZ's exhibit high strength and fracture toughness, studies have shown that annealing at temperatures of $1000^{\circ}C$ or higher (1-4,6) reduces both strength and toughness. It has also been found for Y-TZP that annealing in the temperature range of $150^{\circ}C$ to $400^{\circ}C$ (9-11) has an adverse effect on the microstructure and the subsequent properties. This effect can be accelerated in the presence of water.

The purpose of this paper is to present the results of a study which examined the effect of time and temperature, both high and low, on the properties of Y-TTZ materials.

EXPERIMENTAL PROCEDURE

Billets of Y-TTZ, large enough to have type "B" bend bars (3mm x 4mm x 50mm) machined from them, were obtained from six Japanese manufacturers, Table 1. The bend bars were machined according to Army MIL-STD 1942. Due to material limitations, the AC Sparkplug zirconia was machined into type "A" bars (1.5mm x 2mm x 30mm) according to the same standard. The bulk density of each bar was determined by measuring the mass and geometry. A pulse-echo ultrasonic technique was used to obtain the Modulus of Elasticity (MOE) of the bend bar. The bars, prepared from each of the TTZ's, were then randomly divided into three lots of 30, with each lot to undergo one of the following treatments:

tested as-received
100 hours @ $1000^{\circ}C$
500 hours @ $1000^{\circ}C$

The bars were heat treated, in air at laboratory ambient humidity (40-60%), in an unstressed condition on silicon carbide knife edges to assure uniform heat treatment. The knife edges supported the bend bars well outside the area to be tested during 4-point flexural testing assuring no affect on the subsequent evaluation of mechanical properties. The density and MOE were again measured after heat treatment. The bend bars were then broken at room temperature in 4-point bending, according to MIL-STD 1942, with inner and outer spans of 20 and 40mm respectively and a cross-head speed of 0.5mm/min. (For the type "A" bars, spans of 10 and 20mm were used.) The characteristic strength reported is for the bend bar and has not been corrected for volume and surface effects. Weibull slopes were obtained by interpreting the strength data with a simple least squares curve fit in a stanard weibull two parameter plot. The fracture surface of each bar was examined optically at low magnification to determine the cause of failure. In cases where the cause could not be determined a Scanning Electron Microscope (SEM) was employed and selected fracture surfaces were examined.

TABLE 1. LIST OF TTZ MATERIALS EVALUATED

MATERIAL	PROCESS	MOL% Y_2O_3
COMMERCIAL GRADES		
KYOCERA Z-201	SINTERED	5
TOSHIBA "TASZIC"	SINTERED	2-3
HITACHI	TBD	TBD
NGK LOCKE Z-191	SINTERED	3
KORANSHA "SINTERED"	SINTERED	5
KORANSHA "HIPPED"	HIPPED	5
EXPERIMENTAL GRADES		
AC SPARKPLUG	SINTERED	2.6

TBD - TO BE DETERMINED

Phase analysis of the zirconias was completed using CuK $\propto$ radiation by X-ray diffraction. The ratio of the monoclinic (111) and (11$\bar{1}$) peaks to the tetragonal (101) plus cubic (111) peaks was used to determine the fraction of monoclinic phase present. (2)

Hardness was determined through the Knoop indentation method using a 300g load. Samples were polished on a lead lap using 6-12 um diamond paste.

Also fifteen bars from each TTZ were treated in an autoclave for 50 hours at 200°C and $\sim$.8 MPa water vapor pressure.

RESULTS AND DISCUSSION

High tempeature heat treatments

The experimental results, Table 2 & 3 and Figures 1 & 2, suggest that the seven zirconias be divided into three groups according to their strength and ability to retain that strength after exposure to 1000°C.

- GROUP I: Zirconias with excellent as-received strength and strength retention;

- GROUP II: Zirconias with good as-received strength and good strength retention;

- GROUP III: Zirconias with good as-received strength and poor strength retention.

TABLE 2. SUMMARY OF ZIRCONIA DATA

PROPERTY		MATERIAL UNITS	KY	AC	TOSH	HIT	KS	KH	NGK
DENSITY:	COMPANY LISTING	g/cc	5.9	NDA	6.05	6.08	6.05	NDA	5.91
	AS-RECEIVED		5.853	5.840	5.880	6.038	5.966	6.045	5.869
	100 HRS @ 1000°C		5.803	5.835	5.884	6.029	5.967	6.056	5.861
	500 HRS @ 1000°C		5.772	5.863	5.877	6.037	5.967	6.064	5.863
SONIC MOE:	COMPANY LISTING	GPa	206	NDA	180	209	NDA	NDA	205
	AS-RECEIVED		201	204	200	213	210	214	208
	100 HRS @ 1000°C		203	206	200	213	211	212	207
	500 HRS @ 1000°C		205	208	200	214	210	213	208
MOR (4-PT): (CHARAC. STRENGTH OF BEND BAR)	COMPANY LISTING	MPa	980a	NDA	900a	1000	1100	NDA	1020
	AS-RECEIVED		745	753	633	1169	640	1261	873
	100 HRS @ 1000°C		470	683	581	1053	600	1070	754
	500 HRS @ 1000°C		334	671	576	1062	663	1045	754
WEIBULL NO.:	COMPANY LISTING	NONE	NDA	NDA	NDA	NDA	10.2	NDA	NDA
	AS-RECEIVED		8.8	12.2	6.2	3.6	9.5	8.8	15.2
	100 HRS @ 1000°C		2.3	18.7	14.2	5.9	8.9	5.2	10.0
	500 HRS @ 1000°C		3.0	5.2	13.8	5.0	4.0	12.5	10.6
MOR (4-PT): (MEAN)	COMPANY LISTING	MPa	NDA	NDA	NDA	NDA	NDA	NDA	NDA
	AS-RECEIVED		704	722	587	1045	608	1192	884
	100 HRS @ 1000°C		413	664	561	975	567	974	718
	500 HRS @ 1000°C		296	613	555	974	589	1003	719
STANDARD DEVIATION:	COMPANY LISTING	MPa	NDA	NDA	NDA	NDA	NDA	NDA	NDA
	AS-RECEIVED		75	70	96	265	75	140	65
	100 HRS @ 1000°C		183	42	47	181	76	132	83
	500 HRS @ 1000°C		122	101	48	215	98	89	80
HARDNESS: (KNOOP) (300g LOAD)	COMPANY LISTING	GPa	12.3b	NDA	11.8c	13.3b	14.7c	NDA	11.7
	AS-RECEIVED		10.5	11.1	10.1	12.4	10.8		10.9
	100 HRS @ 1000°C		9.4	11.1	10.0	12.2	11.3		11.3
	500 HRS @ 1000°C		9.4	11.0	10.4	12.2	11.0		10.4

KY - KYOCERA Z201
AC - AC SPARKPLUG
TOSH - TOSHIBA "TASZIC"
HIT - HITACHI

KS - KORANSHA "SINTERED"
KH - KORANSHA "HIPPED"
NGK - NGK-LOCKE Z191

NDA - NO DATA AVAILABLE
a - BELIEVED TO BE 3-PT BEND RESULTS
b - VICKERS 500g LOAD
c - VICKERS

TABLE 3. % MONOCLINIC PHASE CONTENT WITH HEAT TREATMENT TIME

MATERIAL:	AS-RECEIVED % MONO	% T+C	100 HRS @ 1000°C % MONO	% T+C	500 HRS @ 1000°C % MONO	% T+C
Kyocera Z-201	24.5	75.5	55.0	45.0	65.4	34.6
AC Sparkplug	30.0	70.0	38.3	61.7	29.4	70.6
Toshiba "TASZIC"	32.4	67.6	26.7	73.3	44.0	56.0
Hitachi	10.5	89.5	2.4	97.6	20.1	79.9

Figure 1. Room Temperature Strength vs Heat Treatment Time

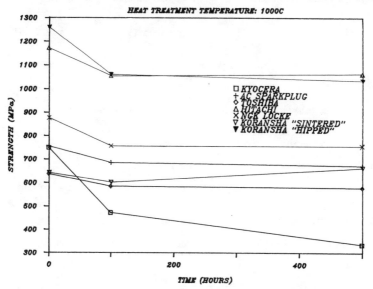

Figure 2. Density vs Heat Treatment Time

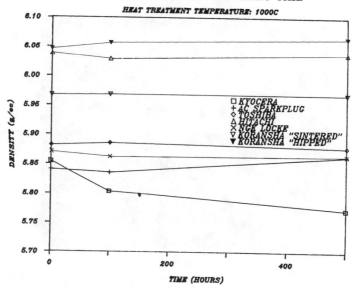

GROUP I includes Hitachi (HIT) and the hipped grade from Koransha (KH). Both exhibit excellent as-received strength, 1169 MPa and 1261 MPa, respectively, and a slight strength loss after 100 hours at 1000°C, but no further loss after 500 hours at 1000°C, Figure 1. However, the HIT strength data is widely scattered as indicated by weibull slopes below 6 for all three cases. The density remains constant for both throughout testing, Figure 2. The monoclinic phase content (%m) is low for HIT and remains essentially low, Table 3 (%m has not been determined for KH).

Although this study has not yet measured grain size, it is believed to be the major contributor to the strength and strength retaining ability of these TTZ's. Lange (12) reports that there is a critical grain size, to prevent spontaneous t→m transformation during cooling, for each TTZ. For grain sizes below this critical value, the thermodynamic free energy necessary for the transformation is increased and few grains transform. Alternatively, when the grains are larger than the critical size, the material can no longer constrain the tetragonal grain in its metastable state, resulting in the t→m transformation. The data supplied by the company lists HIT as having a mean grain size of 0.3um. If this is below the critical value, it would account for the low %m and in turn the high strength, strength retention and constant density of the HIT throughout the testing.

Optical fractography of the fracture surfaces of HIT and KH show that a high percentage (~20%) of failures are due to subsurface flaws rather than surface flaws. Figure 3 is an SEM photo of the most common cause of failure in HIT. It looks like a discrete void, ~20um in size, which was formed when the material was in its green state.

Figure 3. SEM Fractograph of a Discrete Void in a HIT Sample. Flaw Size is Approximately 20 um

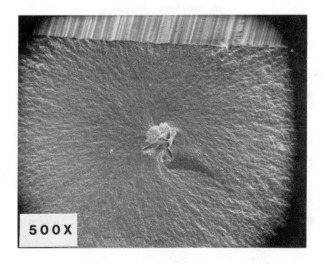

500X

GROUP II includes the AC Sparkplug (AC), Toshiba "TASZIC" (TOSH), NGK-Locke Z-191 (NGK) and the sintered grade from Koransha (KS). These TTZ's have good as-received strengths, but their values are well below that of the Group I. They also exhibit strength retention similar to Group I, a slight loss after 100 hours but virtually no further loss after 500 hours, Figure *. The NGK is the best of this group, having an as-received strength of 873 MPa, dropping to 754 MPa after 100 hours with no further loss after 500 hours. Previous work (1) completed on NGK-Locke Z-191 agrees very well with the data reported here. The %m for AC and TOSH (it has not been determined for NGK and KS) stays constant but is about 3 times higher than HIT. This would account for the lower strength of Group II TTZ's. Optical fractography indicates that the common cause of failure for these TTZ's is either a sintering agglomerate or a pore.

GROUP III includes the Kyocera Z-201 (KY). This TTZ has a good as-received strength but very little strength is retained after heat treatments. The strength loss is 40% after 100 hours and 60% after 500 hours. The density shows a corresponding decrease while %m shows a large increase. The increase in %m accounts for the decrease in density and strength, thus a greatly reduced amount of transformation toughening. The KY exhibits property degradation similar to the best Mg-PSZ examined by Larsen and Adams (1), Schioler (2) and Schioler, et al. (3)

As can be seen from Table 2 that the MOE and hardness of each TTZ remains constant throughout testing except for the KY which lost some of its hardness. However, these properties varied from one TTZ to the next.

Comparison of KS and KH

In addition, an interesting side note is the difference in properties between a hipped and sintered TTZ. Presumably the same material, the KH and KS TTZ's have radically different properties. The density and MOE are slightly higher for the KH and it has almost twice the as-received room temperature strength of the KS. This shows that hipping can improve the properties of a TTZ when compared to the same material sintered, at least in this case, but the full extent of this effect cannot yet be determined.

Additional high temperature heat treatments

Because Group I exhibited excellent properties after heat treatment at 1000°C it was decided that these TTZ's would undergo additional heat treatments to determine if strength retention is maintained at higher temperatures. Two groups of 15 bars were heat treated for 500 hours at either 1100°C or 1200°C. Density and MOE were measured before and after heat treatment. Once heat treated, the bars were broken in 4-point bending, Table 4. Both TTZ's showed no further strength reduction after 500 hours at 1100°C and HIT no further reduction after 500 hours at 1200°C. The KH is now undergoing the 1200°C heat treatment. As with the previous heat treatments density and MOE continued to remain constant.

TABLE 4

STRENGTH DATA FOR HIT AND KH AFTER ADDITIONAL HEAT TREATMENTS

MATERIAL	HITACHI	KORANSHA "HIPPED"
CONDITION		
As-Received MOR (MPa)	1169	1261
StD	265	140
500 Hours @ 1000°C MOR (MPa)	1062	1045
StD	215	89
500 Hours @ 1100°C MOR (MPa)*	1170	1135
StD	130	102
500 Hours @ 1200°C MOR (MPa)*	1098	
StD	74	

* Sample size = 15

Low temperature heat treatments

Figure 4, shows the effects of low temperature annealing and water on the strength of three Y-TTZ's. In all cases there was a significant loss of strength. The HIT exhibited a tremendous 68% loss while AC and KY had losses of 22 and 50%, respectively. A fourth TTZ, TOSH, was also exposed to these conditions but none of the fifteen samples remained intact for 4-point flexure testing. All TTZ's had excessive macrocracking of the surface, which was the direct cause of the strength loss. This macrocracking can be attributed to the tetragonal to monoclinic phase transformation. (8) Similar tests will be carried out on the three remaining TTZ's.

CONCLUSIONS

1) The seven TTZ's evaluated can be divided into three general groups according to their as-received room temperature strength and ability to retain that strength.

2) The behavior of HIT and KH Y-TTZ's demonstrated that it is possible to develop TTZ's which retain up to 90% of their room temperature strength after long time exposures at 1000 - 1200°C.

3) Of the four TTZ's examined, all exhibited strength decreases after being exposed to 200°C and ~0.8 MPa water vapor pressure for 50 hours.

Figure 4. Room Temperature Strength vs. Heat Treat Time after 50 hrs Exposure to 200°C and ～ .8MPa water vapor pressure. Numbers in Parenthesis is the Number of Bars Broken.

ROOM TEMPERATURE STRENGTH VS HEAT TREATMENT TIME

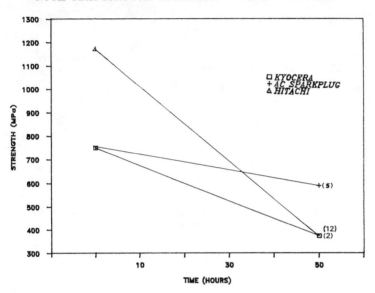

REFERENCES

1. D. C. Larsen and J. W. Adams, "Long-Term Stability and Properties of Zirconia Ceramics for Heavy Duty Diesel Engine Components," prepared for NASA-Lewis Research Center, for US Department of Energy under Contract DEN 3-305 NASA CR-174943, (Sept 1985).

2. L. J. Schioler, "Advanced Transformation Toughened Oxides," prepared for US Department of Energy under Interagency Agreement DE-AI05-840R21411, (unpublished).

3. L. J. Schioler, G. D. Quinn, and R. N. Katz, "Time-Temperature Dependence of the Strength of Commercial Zirconia Ceramics," AMMRC TR 84-16, prepared for U.S. Department of Energy under Interagency Agreement DE-AE-101-77 CS51017, (April 1984).

4. N. L. Hecht, D.E. McCullum, D. W. Grant, J. D. Wolf, G. A. Graves, and S. Goodrich, "The Experimental Evaluation of Environmental Effects in Toughened Ceramics for Advanced Heat Engines," Proceedings of the 23rd Automotive Technology Development Contractors' Coordination Meeting, p. 299, Society of Automotive Engineers, Warrendale, PA, March 1986.

5. M. K. Ferber, and T. Hine, "Time-Dependent Mechanical Behavior of Partially Stabilized Zirconia for Diesel Engine Applications," IBID, p. 285.

6. T. Masaki, "Mechanical Properties of Y_2O_3-Stabilized Tetragonal Polycrstals after Ageing at High Temperatures," J. Am. Ceram. Soc., 69 [7], p. 519, (1986).

7. K. Tsukma, "Mechanical Properties and Thermal Stability of CeO_2 Containing Tetragonal Zirconia Polycrystals," Am. Ceram. Soc. Bull., 65 [10], p. 1386, (1986).

8. M. Ruhle and A. H. Heuer, "Phase Transformation in ZrO_2-Containing Ceramics: II, The Martensitic Reaction in t-ZrO_2," p. 14, Advances in Ceramics, Vol. 12, Science and Technology of Zirconia, ed. N. Claussen, M. Ruhle, and A. H. Heuer. The American Ceramic Society, (1984).

9. T. Sato and M. Shimada, "Crystalline Phase Change in Yttria-Partially-Stabilized Zirconia by Low-Temperature Annealing," Comm. Am. Ceram. Soc., p. c212, (1984).

10. T. Sato and M. Shimada, "Transformation of Yttria-doped Tetragonal ZrO_2 Polycrystals by Annealing in Water," J. Am. Ceram. Soc., 68 [6] p. 356, (1985).

11. T. Sato, S. Ohtaki and M. Shimada, "Transformation of Yttria Partially Stabilized Zirconia by Low Temperature Annealing in Air," J. Mat. Sci., [20] p. 1466 (1985).

12. F. Lange, "Transformation Toughening - Part 2," J. Mat. Sci., [17] p. 225, (1982).

MECHANICAL PROPERTIES AND THERMAL STABILITY OF YTTRIA-DOPED TETRAGONAL ZIRCONIA POLYCRYSTALS WITH DIFFUSED CERIA IN THE SURFACE

T. SATO, S. OHTAKI, T. FUKUSHIMA, T. ENDO AND M. SHIMADA
Department of Applied Chemistry, Faculty of Engineering, Tohoku University,
Sendai 980, Japan

ABSTRACT

Yttria-doped tetragonal zirconia polycrystals (Y-TZP) containing 2 and 3 mol% of Y_2O_3 uniformly alloyed by 0-15 mol% of CeO_2 or diffusing CeO_2 on the surface were fabricated, and the mechanical properties and thermal stability of the sintered bodies were evaluated by annealing in humidity conditions at 50-600°C. The tetragonal-to-monoclinic phase transformation proceeded at 100-500°C in air, and accompanied microcracks. The phase transformation proceeded rapidly on the surface, but slowly inside the body. The bending strength of the annealed specimens depended on the depth of the transformation layer thickness, but not on the degree of the phase transformation on the surface. Alloying CeO_2 was useful to improve the thermal stability of Y-TZP, but noticeably decreased the fracture strength. Diffusing CeO_2 on the suface of Y-TZP seemed to be useful to improve the thermal stability without loss of the fracture strength.

INTRODUCTION

Many studies have been focused on the structural application of Y-TZP because of the superior mechanical properties and the high sinterabilty [1,2]. However, the strength and toughness of Y-TZP sometimes greatly degrade by annealing at low temperature such as 150-300°C in air [3-6]. It was believed to be due to the formation of microcracks accompanied with the tetragonal-to-monoclinic phase transformation during annealing. In the previous study, it was reported that the phase transformation was greatly accelerated by water molecules [4,5], and that the thermal stability of Y-TZP could be improved by microstructural modification to increase the free energy change of the tetragonal-to-monoclinic phase transformation [6], which can be written by

$$G_{t \to m} = \Delta G_c + \Delta G_{se} + \Delta G_s \qquad (1)$$

Mat. Res. Soc. Symp. Proc. Vol. 78. ©1987 Materials Research Society where ΔG_c, ΔG_{se} and ΔG_s are the chemical free energy change, the strain free energy change and the surface free energy change of the phase transformation. ΔG_c may be controlled by doping stabilizers such as CaO, MgO, CeO_2 and TiO_2. CeO_2 is a good candidate to improve the thermal stability of Y-TZP because it can greatly decrease the critical transformation temperature. However, it was pointed out that the sinterability of CeO_2-TZP was not good, and that the fracture strength of CeO_2-TZP was not high, because the tetragonal-to-monoclinic phase transformation proceeds at relatively low stress field. Since the tetragonal-to-monoclinic phase tranformation proceeds from the surface, it is expected that the thermal stability of Y-TZP may be improved without loss of the fracture strength by doping CeO_2 on the surface of thin layer less than the critical flaw length. In the present study, Y-TZP uniformly alloyed by 0-15 mol% of CeO_2 and diffusing CeO_2 on the surface were fabricated and annealed in the controlled humidity atmosphere to clarify the possibility to improve the thermal stability and mechanical properties.

EXPERIMENTAL

Zirconia powders containing 2 and 3 mol% Y_2O_3 were mixed with a weighed amount of CeO_2 powder by ball-milling with acetone and plastic balls in a plastic container, then all powders were dried. These powders were cold isostatically pressed at 200 MPa to form plates, 5 by 30 by 50 mm, and then were sintered at 1500 and 1600°C for 3 hr in air. The sintered plates were cut into rectangular coupons, 2 by 4 by 12 mm and polished to a mirror-like surface to measure the fracture strength. For diffusing CeO_2 on the surface of Y-TZP, the polished specimens were molded by CeO_2 powder and pressed by 50 MPa, and then calcined at 1400°C for 2-10 hr. The specimens obtained in these ways were annealed in controlled humidity atmosphere and water regulated at a desired temperature. The phase identification was carried out by X-ray diffraction analysis. The amount of the monoclinic (m) ZrO_2 contentet was determined by Garvie's method [7]. The bulk density of a sintered body was measured by Archimedes technique. Microstructure of a specimen was observed by scanning electron microscopy. The average grain size was determined by intercept method [8]. The CeO_2 content doped on the surface was determined by electron probe micro-analysis (EPMA). Fracture strength was determined by 3-point bending test with cross head speed of 0.5 mm/min.

RESULTS AND DISCUSSION

Annealing of Y-TZP in Controlled Humidity Atmosphere

Y-TZP containing 2 mol% Y_2O_3 was entirely tetragonal phase, and that contaning 3 mol% Y_2O_3 was the mixture of the tetragonal and cubic phases. No m-ZrO_2 was observed in these as-sintered materials. Amount of m-ZrO_2 formed on the surface and the bending strength of ZrO_2-2 mol% Y_2O_3 ceramics of different grain size annealed in air at 50-500°C and 3.35 kPa of water vapor pressure for 50 hr is shown in Fig. 1. The tetragonal-to-monoclinic phase transformation proceeded above 100°C and the bending strength greatly decreased around 200°C. It was noticeable that even though the degree of the phase transformation on the surface was almost the same, the degree of the degradation of the bending strength was quite different.

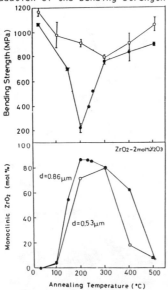

Fig. 1 Changes in the bending strength and the crystalline phase composition on the surface by annealing at various temperatures.

Time dependence of the bending strength of ZrO_2-2 mol% Y_2O_3 and ZrO_2-3 mol% Y_2O_3 is shown in Fig. 2. The tetragonal-to-monoclinic phase transformation on the surface completed within 50 hr. The bending strength of ZrO_2-2 mol% Y_2O_3 and ZrO_2-3 mol% Y_2O_3 having the grain size of 0.86 and 1.83 µm, respectively , decreased less than 100 Mpa within 100 hr, but ZrO_2-3 mol% Y_2O_3 having the grain size of 0.81 µm possessed the bending strength of more than 800 MPa even after 1000 hr annealing. These results indicated that the degree of the degradation of the fracture strength was greatly depended on both the grain size of zirconia and Y_2O_3 content.

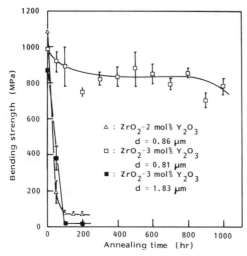

Fig. 2 Time dependence of the bending strength of Y-TZP annealed at 200°C and 3.35 kPa of water vapor pressure.

The scanning electron microgrphs of the cut surface of ZrO_2-2 mol% Y_2O_3 with the different grain size of zirconia annealed at 200°C for 50 hr are shown in Fig. 3. Microstructural change caused by annealing was observed on the surface but not inside the body. The transformation layer thickness in ZrO_2-2 mol% Y_2O_3 with the grain size of 0.86 and 0.53 μm were 410 μm and 26.7 μm, respectively.

The relation between the bending strength of the annealed specimen, σ_b, and the transformation layer thickness, C, is shown in Fig. 4. The bending strength was almost constant until the transformation layer thickness exceeded 50 μm, and then decreased with increasing the transformation layer thickness.

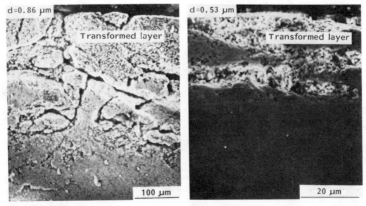

Fig. 3 Scanning electron micrographs of the cut surface of Y-TZP annealed at 200°C and 3.35 kPa of water vapor pressure for 100 hr.

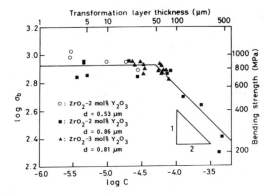

Fig. 4   Relation between the bending strength and the transformation layer thickness of Y-TZP annealed.

By applying the small scale yielding approximation, the stress field around the crack tip is given by

$$\sigma_r = K_{IC} f(\theta)/\sqrt{2\pi r} \qquad (2)$$

where K_{IC} is the crack tip stress intensity factor, r is the distance from the crack tip and $f(\theta)$ is the angular dependent term [9]. Since $f(\theta)$ is unity at right angle to the crack plane, following equation can be derived.

$$\log \sigma_r = \log K_{IC}/\sqrt{2\pi} - 1/2 \log r \qquad (3)$$

As seen in Fig. 4, when the transformation layer thickness exceeded 50 µm, the plots of $\log \sigma_b$ versus $\log C$ was fitted to the straight line with a slope of 1/2, which agreeed with the value shown by equation (3). These results indicated that the fracture origin of the annealed sample whose transformation layer thickness was more than 50 µm was the cracks introduced on the surface by the tetragonal-to-monoclinic phase transformation.

Characteristics of Y-TZP Homogeneously Alloyed with CeO_2

In order to prevent the tetragonal-to-monoclinic phase transformation, 0-12 mol% of CeO_2 was added to Y-TZP ceramics containing 0-6 mol% Y_2O_3 and the thermal stability was evaluated by annealing in water at 100°C for 7 days. The relation among the amount of $m-ZrO_2$ formed on the surface, Y_2O_3 content and CeO_2 content in TZP is shown in Fig. 5, where the shaded part indicated the region where the tetragonal-to-monoclinic phase transformation did not proceed at all. The degree of the phase transformation decreased

152

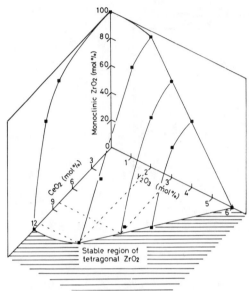

Fig. 5 Relation among the concentration of Y_2O_3 and CeO_2 and the amount of m-ZrO_2 formed by annealing in water at 100°C for 7 days.

with increasing the concentration of both Y_2O_3 and CeO_2, i.e. the amount of CeO_2, which could completely control the phase transformation, decreased from 12 to 0 mol% with increasing the Y_2O_3 content from 0 to 6 mol%. Since the mechanical properties of Y-TZP greatly degrade with decreasing Y_2O_3 content below 2 mol% and increasing above 3 mol%, Y_2O_3 content should keep by between 2 and 3 mol%. As seen in Fig. 5, it is necessary to alloy 8-11 mol% of CeO_2 with Y-TZP containing 2-3 mol% Y_2O_3 in order to prevent the phase transformation perfectly.

The bending strength and the relative density of Y-TZP alloyed with various amount of CeO_2 are shown in Fig. 6. Both the bending strength and

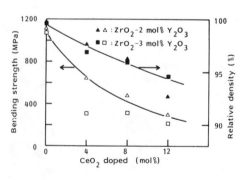

Fig. 6 Bending strength and relative density of Y-TZP alloyed with various amount of CeO2.

the relative density greatly decreased with increasing the CeO_2 content. The loss of the bending strength might be due to the increase in the flaw size in as-sintered materials. These results indicated that alloying Y-TZP with CeO_2 to improve the thermal stability degraded the sinterability and the fracture strength.

Characteristics of Y-TZP Diffused CeO_2 on the Suface.

In order to improve the thermal stability of Y-TZP without loss of the mechanical properties, CeO_2 was diffused on the surface of Y-TZP by calcining Y-TZP in CeO_2 powder bed at 1400°C for various time. CeO_2 contents diffused on the surface by calcining the sintered bodies of ZrO_2-2 mol% Y_2O_3 and ZrO_2-3 mol% Y_2O_3 in CeO_2 powder bed at 1400°C for 10 hr are determined by EPMA. The results are shown in Fig. 7. About 10 mol% of CeO_2 was doped on the surface of both specimens. CeO_2 content decreased with increasing the distance from the surface.

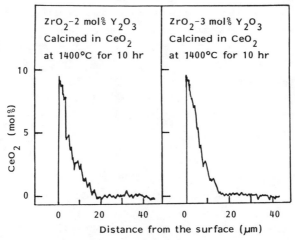

Fig. 7   Relation between the CeO_2 content diffused in Y-TZP and distance from the surface.

The specimens of ZrO_2-2 mol% Y_2O_3 calcined in CeO_2 powder bed at 1400°C for various times were annealed in air at 200°C and 3.35 kPa of water vapor pressure for 100 hr to evaluate the thermal stability. The bending strength of the specimens both before and after annealing and the amount of m-ZrO_2 formed on the surface by annealing are shown in Fig. 8. The amount

154

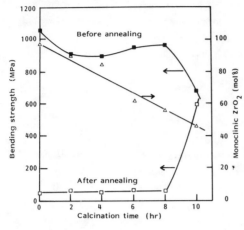

Fig. 8　Bending strength and amount of m-ZrO_2 formed by 100 hr annealing at $200°C$ of ZrO_2-2 mol% Y_2O_3 diffused CeO_2 on the surface.

of m-ZrO_2 formed decreased with increasing the calcination time. The bending strength before annealing decreased from 1060 MPa to 680 MPa by increasing the calcination time from 0 to 10 hr. The bending strength after annealing of the specimens calcined less than 8 hr decreased to below 100 MPa, but the specimen calcined for 10 hr still possessed the bending strength of 600 MPa after annealing. The loss of the bending strength by diffusing CeO_2 on the surface was much smaller than that by alloying CeO_2 homogeneously as shown in Figs. 6 and 8. The present results indicated the possibility that the thermal stability of Y-TZP can be improved without loss of the mechanical properties by doping CeO_2 on the surface.

References

1. R.C. Garvie, R.H.J. Hannink, T.T. Pascoe, Nature (London), 258(5337), 703 (1975).

2. D.L. Porter, A.H. Heuer, J. Am. Ceram. Sci., 60(3-4), 183 (1977).

3. K. Kobayashi, H. Kuwajima, T. Masaki, Solid State Ion., 3-4, 489 (1981).

4. T. Sato, M. Shimada, J. Am. Ceram. Soc., 68(6), 356 (1985).

5. T. Sato, S. Ohtaki, T. Endo, M. Shimada, J. Am. Ceram. Soc., 68(12), C320 (1985).

6. T. Sato, M. Shimada, J. Mater. Sci., 20, 3988 (1985).

7. R.C. Garvie, P.S. Nicholson, J. Am. Ceram. Soc., 55(6), 303 (1974).

8. R.L. Fullman, J. Metals Trans. AIME, 197(3), 447 (1953).

9. H.V. Swain, Acta Metall., 33, 2083 (1985).

STRENGTH IMPROVEMENT IN TRANSFORMATION TOUGHENED CERAMICS
USING COMPRESSIVE RESIDUAL SURFACE STRESSES

R. A. CUTLER,* J. J. HANSEN,* A. V. VIRKAR,** D. K. SHETTY,** AND
R. C. WINTERTON,***
* Ceramatec, Inc., 2455 S. 900 W., Salt Lake City, UT 84119
** Department of Materials Science and Engineering. University
of Utah, Salt Lake City, UT 84112
*** The Dow Chemical Company, Central Research Inorganic Materials
and Catalysis Laboratory, Midland, MI 48674

ABSTRACT

Al_2O_3-15 vol. % ZrO_2 bar shaped composite specimens were
fabricated by pressing three layers. The two outer layers consisted
of Al_2O_3 and unstabilized ZrO_2 (primarily in the monoclinic
polymorph), and the inner layer consisted of Al_2O_3 and partially
stabilized zirconia in the tetragonal polymorph. The transformation
of ZrO_2 from tetragonal to monoclinic, upon cooling from sintering
temperature, led to the establishment of residual compressive
stresses in the outer layers. Flexural tests at room temperature
showed that residual stresses contributed to strength increasing
from 450 to 825 MPa. The existence of these stresses was verified
by measuring apparent fracture toughness, as well as using strain
gages. Strength and toughness data were obtained at 500, 750,
and 1000°C. X-ray diffraction was used to explain the elevated
temperature data by monitoring the monoclinic to tetragonal
transformation upon heating to 1000°C.

INTRODUCTION

Residual stresses have been used to strengthen a number
of ceramic or ceramic-metal composites including glass[1], glass-
ceramics[2], ZrO_2[3-6], Si_3N_4-SiC[7], and WC-Co[8]. Significant
strength increases occur when the applied tensile load must
first overcome the residual compressive stresses near the surface
of the component. There may be a strength decrease if uniform
tension were applied to the specimen (provided failure occurs
from interior flaws) since there are balancing tensile residual
stresses in the interior of the specimen. Since surface grinding
often introduces larger flaws than occur during processing,
strength improvement may occur from compressive surface stresses
even if the component is loaded in uniform tension. Surface
compressive stresses should also improve the strength of ceramics
loaded in compression since failure often occurs due to tensile
contact stresses. Applications, such as the adiabatic diesel
engine, would benefit from ceramics able to withstand contact
loading and thermal stresses. Transformation toughened ceramics,
with superior strength and toughness at ambient temperature,
lose much of their strength and toughness upon heating to 750°C.
Previous approaches relied upon grinding[3-5] or high temperature diffusion[6] to introduce compressive stresses in surface
layers that were limited to depths of less than 50 microns.
An alternate method for establishing residual stresses in transformation toughened ceramics has been developed by Virkar et al.[9,10].
The depth of compressive stresses is controlled during green
processing. Surface zones of compressive stress exceeding 200

microns can be established to contain surface flaws well within the zone, thus leading to increased resistance to crack propagation. Furthermore, transformation-induced stresses created upon cooling from the sintering temperature lead to strength increases at 25°C in excess of 350 MPa without grinding or post-sintering heat treatments.

The purpose of this paper is to report data for Al_2O_3-ZrO_2 ceramics which demonstrate substantial improvement in strength and apparent toughness at ambient temperature due to transformation-induced stresses. Strength and apparent toughness enhancements were observed at temperatures up to 750°C. X-ray diffraction was used to explain strength degradation of composites with transformation-induced stresses at higher temperatures.

EXPERIMENTAL PROCEDURE

Al_2O_3-15 vol. % ZrO_2 (partially stabilized with 2 mole % Y_2O_3) and Al_2O_3-15 vol. % ZrO_2 (no zirconia stabilizer added) powders[1] were dispersed, vibratory milled, and spray dried. The dried powders were screened through a 170 mesh screen before the monolithic and the three layered Al_2O_3-ZrO_2 composite bar specimens were fabricated. Three layer composites were made by loading a predetermined volume of "outer layer" ($Al_2O_3ZrO_2$(unstabilized)) powder into a steel die, followed by "inner layer" (Al_2O_3-$ZrO_2(Y_2O_3)$) powder and a second addition of "outer layer" powder. Outer layer thicknesses of three sets of Al_2O_3-15ZrO_2 bars were nominally 375, 750, and 1500 micrometers after sintering. Monolithic bars were also made of the inner and the outer layer compositions. The composites were uniaxially pressed at 35 MPa, followed by isostatic pressing at 207 MPa. The bars (4.5 mm x 5 mm x 50 mm after sintering) were chamfered in the green state (or after bisquing at 800°C) and were not subjected to grinding operations after sintering. The bars were sintered at 1587°C for 60 minutes and hot isostatically pressed (HIP) at 1500°C and 175 MPa for 30 minutes in Ar overpressure.

Fracture strengths were assessed in four-point bending (20 mm inner span and 40 mm outer span) at a crosshead speed of 0.5 mm/min. X-ray diffraction (XRD) was used to determine the relative amounts of the ZrO_2 polymorphs[11] and these data were used to calculate the theoretical density. A high temperature x-ray diffractometer with a long arm goniometer[2] was used in the high temperature phase study. Three specimens each of the monolithic "outer layer" composite and the three layer composite having an outer layer thickness of 375 microns were polished on one side to a 1 micron finish. Hardness and apparent fracture toughness by a multiple-flaw indentation technique[12] were measured at 25, 500, 750, and 1000°C. Apparent toughness was also measured by indentation[13] at room temperature on bars which were in the "as-sintered" condition (i.e., no grinding or polishing). Residual stresses were measured using a strain gage technique described previously[9].

---

[1] Reynolds HP-DBM (Reynolds Metal Co., Bauxite, AK) alumina was mixed with DK-1 (Daiichi Kagaku Kogyo Co., Ltd., Osaka, Japan) zirconia for the outer layers and Zircar 2.0 mole % Y_2O_3 doped ZrO_2 (Zircar Corp., Florida, NY) for the inner layer.

[2] Model HTK-10 (Anton Parr, Austria).

RESULTS AND DISCUSSION

Densities of the as-sintered specimens were between 97.3 and 97.6 % of theoretical which increased to greater than 99.8% of theoretical upon hot isostatic pressing. Shrinkage was identical for the inner and the outer layers. Zirconia in the monolithic "inner layer" specimens was all tetragonal, while the monoclinic content of the monolithic "outer layer" specimens was 64%. The monoclinic content in the three layer composites was measured as 64%, 60%, and 49% for outer layer thicknesses of 1500, 750, and 375 microns, respectively. The decrease in monoclinic content as a function of decreasing outer layer thickness is believed to be due to the increased constraint of the matrix due to the presence of the residual stresses. Hardness was independent of the outer layer thickness, d_1, as expected, and ranged between 17 and 18 GPa.

Strength increased with decreasing outer layer thickness due to the increasing residual compressive surface stresses (see Figure 1). Fractography showed that majority of the failures initiated from the tensile surface, as opposed to previous results which showed failure from near the interface of the three layer composites due to inclusions or voids[9]. The magnitude of the compressive stress, σ_c, assuming a square wave distribution and no difference in moduli or thermal expansion between the inner and outer layers, is given[9] by

$$\sigma_c = -d_2 E \Delta\varepsilon_o / d(1-\nu) \tag{1}$$

where $\Delta\varepsilon_o$ is the elastic strain imposed on the outer layers by the constraint of the inner layer[9]. E is Young's modulus, ν is Poisson's ratio, d_2 is the inner layer thickness, and d is the bar thickness. For fracture initiating from the outer surface, the predicted failure stress, σ_f, is given by

$$\sigma_f = \sigma_f{}^o + d_2 E \Delta\varepsilon_o / d(1-\nu) \tag{2}$$

where $\sigma_f{}^o$ is the failure stress in the absence of residual stress. This equation assumes that the amount of the monoclinic phase is the same regardless of the outer layer thickness. XRD, however, shows that this is not true due to the differences in the constraint factor. A plot of the experimentally determined failure stresses as a function of d_2/d (see Figure 1) gives a near linear fit of the data ($r^2 = 0.975$), as predicted by Equation (2). The slope of the line (assuming E to be 365 GPa and ν as 0.25) yields a value of $\Delta\varepsilon_o$ as 8.9×10^{-4}. Assuming the linear expansion of t --> m ZrO_2 as 1/3 the volume expansion (taken as 0.049[14]), the percent monoclinic in the outer layer is calculated to be 36%, within a factor of two of the value of 49 to 64% measured by XRD. XRD measures the monoclinic content in the near surface regions. It is expected that this would be higher than that in the interior due to the fact that there is very little constraint in the near surface regions which effectively increases the monoclinic content. The strain gage is therefore a better technique for determining surface compressive stresses[9].

Based on previous strength data[9] which showed decreasing strength as a function of increasing d_2/d, due to internal inclusions and voids, it is likely that the failure initiation site can be controlled by changing the magnitude of $\Delta\varepsilon_o$. If $\Delta\varepsilon_o$ is small, failure will occur from the surface and thinner outer layer thickness is desirable to maximize the compressive stresses.

158

On the other hand, if $\Delta\varepsilon_0$ is large, failure will occur from the interface and it is better to have the outer layer thickness large.

Using a two parameter Weibull distribution, failure probabilities for the monolithic and three layer composites are shown in Figure 2. The lowest strength specimen of the 375 micron outer layer bars had strength greater than the highest strengths of the monolithic bars (see Figure 2). The increase in Weibull modulus (16.1 for three layer specimens with d_1 = 375 micrometers as compared to 9.9 for monolithic "outer layer" specimens) can be explained by superposition of stresses[10]. The surface compressive stresses effectively shift the strength distribution to higher strength values. A three parameter Weibull expression,

$$F = 1 - \exp[-[(\sigma_f-\sigma_u)/\sigma_o]^m] \qquad (3)$$

where F is the fracture probability, σ_u is the "threshold stress" (i.e., the minimum stress for which fracture can occur), σ_o is a scale parameter and m is the Weibull modulus, was used to examine the change in σ_u as a function of compressive stresses (i.e., d_2/d). The "threshold stress" was increased from zero in an iterative process to maximize the coefficient of determination (r^2). The threshold stress was taken as the stress at which r^2 was a maximum[15] (see Figure 3). Weibull plots (see Figure 4) showed that the three parameter function gave a better fit of the data than the two parameter Weibull distribution. It is interesting to note that the threshold stress for the monolithic outer layer specimen was 370 MPa while that of the layered composite with d_1=375 micron was 660 MPa as shown in Figure 3. This tendency of increasing threshold stress with increasing compressive stress was not observed for the other three layer composites and larger sample populations are necessary before firm conclusions can be drawn. Increasing threshold stress is consistent with increasing strength due to the surface compressive stresses.

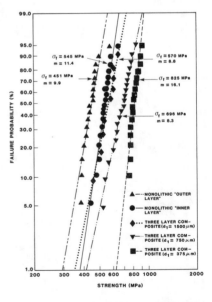

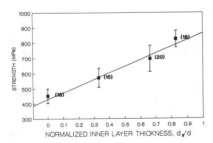

Figure 1. Strength as a function of normalized inner layer thickness, d_2/d (see Equation (2)).

Figure 2. Failure probabilities of monolithic and three layered Al_2O_3-$15ZrO_2$ composites.

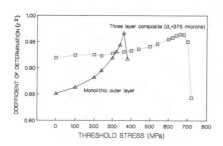

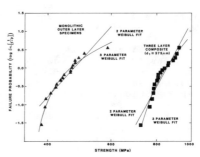

Figure 3. Change in coefficient of determination (r^2) as a function of threshold stress, σ_u.

Figure 4. Comparison of two and three parameter Weibull distributions.

As further evidence of substantial residual stresses in layered composites made by the present technique, strain gauges were attached on fractured strength specimens and the opposite outer faces were successively removed by grinding as reported previously[9]. When the amount of material removed, δ, is small relative to the total thickness of the sample, d, the measured strain, $\varepsilon_M(\delta)$, can be used[9] to approximate $\Delta\varepsilon_o$ by

$$\Delta\varepsilon_o \simeq \varepsilon_M(\delta)d^2/\delta 3d_2 \qquad (4)$$

The slope of the initial linear region of the data displayed in Figure 5 was used to calculate $\Delta\varepsilon_o$, as well as the residual compressive stresses. As expected, the monolithic outer and inner layer materials showed no change in strain upon grinding. The three layered bars, however, showed increasing slopes with increasing residual stresses as shown in Figure 5. The average change in strain, $\Delta\varepsilon_o$ was calculated to be 1.49×10^{-3}, in good agreement with the strength data. The residual stresses, calculated by substituting Equation (4) into Equation (1), were 526, 397, and 317 MPa for the 375, 750, and 1500 micron d_1 specimens, respectively. These are also in good agreement with the change in strength as shown in Figures 1 and 2.

Indentation fracture toughness measurements[13] were made on monolithic and three layer composites in the "as-sintered condition" by breaking strength bars which had been indented at 100N. All bars were assumed to have a hardness of 17 GPa and a modulus of 365 GPa. Fracture toughness is a material property and is not affected by residual stresses. The apparent toughness, K_c^a, for surface cracks contained well within the outer layer region is dependent on residual stresses[10] and is given by

$$K_c^a = K_c^o - \sigma_c 2\sqrt{c}/\sqrt{\pi} \qquad (5)$$

where K_c^o is the fracture toughness (in the absence of residual stresses) and c is the crack radius. The apparent toughness is therefore given by

$$K_c^a = K_c^o + 2\sqrt{c}d_2E\Delta\varepsilon_o/d(1-\nu)\sqrt{\pi} \qquad (6)$$

The apparent toughness as a function of d_2/d is shown in Figure 6 for the monolithic outer layer composites as well as the three layer composites. It should be noted that a "linear" fit of the data ($r^2 = 0.993$) in Figure 6 implies that crack length is a constant. As verified by multiple indent[12] measurements on polished specimens, crack length is not a constant but rather decreases by 15-30% as compressive surface stresses increase. This is observed in Figure 6 by the decreasing K_C^a at higher values of d_2/d. Taking an average value for crack size as 125 microns, the change in strain, $\Delta\varepsilon_0$, as calculated from the slope of the data in Figure 6 is 1.39×10^{-3}, in excellent agreement with the values of 1.47×10^{-3} (calculated from XRD data) and 1.49×10^{-3} (determined via strain gauging). Further work is underway to determine the effect of change in apparent toughness as a function of crack size. Since the apparent toughness is strongly influenced by residual stresses and is expected to increase with increasing crack size, one should be careful as to how the data is used.

Strength data at 25, 500, 750, and 1000°C are shown in Figure 7 for monolithic as well as three layer composites. The change in strength between the three layer composites ($d_1 = 375$ microns) and the monolithic "outer layer" specimens. which is taken as the measure of surface compressive residual stresses, is shown in Figure 8. The residual stresses decrease slightly between 25 and 500°C, more significantly between 500 and 750°C, and rapidly between 750 and 1000°C. X-ray diffraction was used to determine the amount of monoclinic ZrO_2 as a function of temperature on a monolithic outer layer specimen. Figure 9 shows the decrease in monoclinic content for a monolithic "outer layer" specimen as a function of temperature. The percent monoclinic was initially 68.5%, and decreased slightly at temperatures up to 700°C, with rapid conversion to tetragonal ZrO_2 above 700°C (see Figure 9). The x-ray diffraction data are in excellent agreement with the strength change measured experimentally (compare Figures 8 and 9). It should be noted that one would expect monoclinic ZrO_2 in the outside layers of a three layer composite to convert to tetragonal ZrO_2 at lower temperatures that the monolithic specimen due to constraint imposed on the particles by the residual compressive stresses. The present XRD specimen

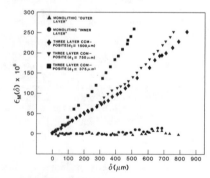

Figure 5. Strain determination in outer layer of Al_2O_3-ZrO_2 bars. Note that monolithic specimens show no indication of residual stresses.

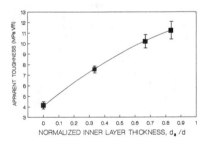

Figure 6. Apparent toughness increase with increasing residual compressive stresses for Al_2O_3-$15ZrO_2$.

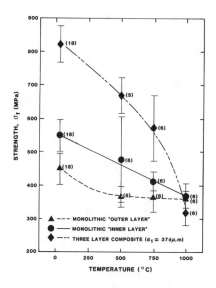

Figure 7. Strength of Al_2O_3-15ZrO_2 as a function of temperature. Note that substantial residual stresses are retained to 750°C.

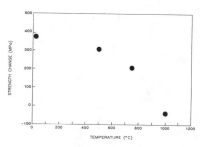

Figure 8. Strength change between three layer composite and monolithic "outer layer" specimens as a function of temperature.

holder does not permit the specimen thickness to exceed 250 microns so three layer composites could not be tested. The observed decrease in strength of the layered composite between 25 and 500°C could alternatively be due to moisture sensitivity or to temperature stress, although thermal expansion data did not correlate with strength data. The important conclusion from the strength data is that significant compressive stresses were retained to 750°C and that strength retention to higher temperature can be expected by controlling ZrO_2 particle size distribution or substituting HfO_2 for ZrO_2.

Changing the size distribution was investigated by using a zirconia powder[3] with a median particle size of 1.35 microns (all particles less than 3.7 microns, 90% less than 2.2 microns, and 90% greater than 0.65 microns). Specimens made with this 1.4 micron powder were compared with samples made with finer ZrO_2 powder (the outer layer zirconia powder used for the data reported herein). The finer ZrO_2 powder contained no particles greater than 1.8 microns and had an average particle size of 0.4 microns (90% less than 1.2 microns and 70% greater than 0.2 microns). Figure 10 shows the percent monoclinic ZrO_2 in the monolithic "outer layer" bars of 15 and 20 vol% 0.4 micron ZrO_2, as well as 15 vol. % 1.4 micron ZrO_2 as a function of temperature. The 1.4 micron ZrO_2 powder, although comparable in monoclinic content to the 0.4 micron ZrO_2 at temperatures up to 800°C, had twice as much monoclinic zirconia at 1000°C. While room temperature strengths of three layer composites of the two materials are comparable, it is predicted that improved retention in strength will occur by using the 1.4 micron zirconia which will not transform to the tetragonal phase until higher temperatures. An alternate method for increasing the ZrO_2 size is to increase the volume percent, allowing particle growth

[3]K906 Tededyne Wah Chang Albany (Albany, OR)

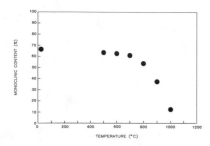
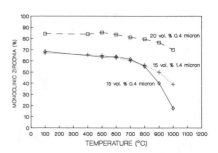

Figure 9. X-ray diffraction data showing the % monoclinic ZrO_2 in a monolithic Al_2O_3-15ZrO_2 (outer layer) specimen as a function of temperature.

Figure 10. X-ray diffraction showing greater retention of monoclinic ZrO_2 by coarsening the ZrO_2 particle size.

during sintering. The data for 20 vol. % ZrO_2, using the 0.4 micron ZrO_2, in Figure 10 indicates that smaller ZrO_2 particles coarsened during sintering leading to substantial improvement in the retention of the monoclinic polytype. Strength measurements as a function of temperature are planned to determine if XRD can be used to predict the retention of strength at temperatures to 1000°C.

Apparent toughness as a function of temperature was determined using the multiple indent technique[12] with monolithic "outer layer" and three layer (d_1=375 microns) composite bars as shown in Figure 11. The data for three layer composites show more scatter than monolithic bars due to the variability (between bars) in the amount of outer layer removed while polishing one side of the bar, causing a change in residual stresses. It is believed that removal of a uniform amount of material from both outer layers would eliminate this tendency. The apparent toughness decreases with increasing temperature in a manner similar to strength data, as expected.

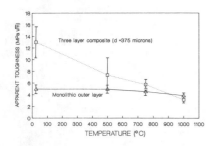

Figure 11. Apparent toughness of Al_2O_3-15ZrO_2 as a function of temperature. Note that substantial residual stresses are present below 750°C.

CONCLUSIONS

The existence of substantial residual stresses in three layer Al_2O_3-15ZrO_2 composites was verified by x-ray diffraction, strain gauge, strength, and apparent toughness measurements. X-ray diffraction measurements were used to explain the decrease in strength and toughness at elevated temperatures. The strength and apparent toughness of layered composites was linearly related to the normalized depth of the inner layer, consistent with failure initiation from the outer surface

of the bars. Composites with improved retention of monoclinic
ZrO_2 content at temperatures above 750°C were identified by
x-ray diffraction. Elevated temperature testing of these composites,
as well as composites where HfO_2 has been substituted for ZrO_2,
will be used in future work to show that transformation toughened
ceramics can be fabricated with excellent strength retention
to 1000°C.

ACKNOWLEDGEMENTS

This work was supported by Oak Ridge National Laboratory
under subcontract 86-X22028C under the Ceramic Technology for
Advanced Heat Engine Project. J. Bright and D. Singh performed
high temperature strength measurements.

REFERENCES

1.   D.G. Holloway, The Physical Properties of Glass, (Wykeham
     Publications Ltd., London, 1973) pp. 190-196.
2.   D.A. Duke, J.E. Megles, J.E. MacDowell and H.E. Boop, J. Am.
     Ceram. Soc., 51, 98-102 (1968).
3.   R.T. Pascoe and R.C. Garvie, in Ceramic Microstructures
     '76 edited by R.M. Fulrath and J.A. Pask (Westview Press,
     Boulder, CO 1977) pp. 774-784.
4.   J.S. Reed and A. Lejus, Mater. Res. Bull., 12 (10), 949-954
     (1977).
5.   N. Claussen and J. Jahn, Ber. Dtsch. Keram. Ges., 55 (11),
     487-491 (1978).
6.   D.J. Green, J. Am. Ceram. Soc., 66 (10), C-178-179 (1983).
7.   M.L. Torti and D.W. Richerson. U. S. Patent No. 3,911.188
     (7 Oct. 1975).
8.   R.A. Cutler and A.V. Virkar, J. Mater. Sci., 20, 3557-3573
     (1985).
9.   A.V. Virkar, J.L. Huang and R.A. Cutler, accepted for publi-
     cation in J. Am. Ceram. Soc.
10. R.A. Cutler, J.B. Bright, A.V. Virkar and D.K. Shetty,
     submitted for publication to Am. Ceram. Soc.
11. H. Toraya, M. Yoshimura and S. Somiya, J. Am. Ceram. Soc.,
     68, C-119-121 (1984).
12. R.F. Cook and B.R. Lawn, J. Am. Ceram. Soc., 66 (11), C-200-
     201 (1983).
13. P. Chantikul, G.R. Anstis, B.R. Lawn and D.B. Marshall,
     J. Am. Ceram. Soc., 64 (9), 539-543 (1981).
14. W.M. Kriven, in Advances in Ceramics, Vol. 11, Science
     and Technology of Zirconia II, Edited by N. Claussen, M. Ruhle,
     and A.H. Heuer (Am. Ceram. Soc., Columbus, OH, 1984) pp. 64-77.
15. D.K. Shetty, unpublished work.

ZIRCONIA DISPERSED MULLITE CERAMICS THROUGH HOT-PRESSING OF AMORPHOUS ZrO_2-SiO_2-Al_2O_3 OBTAINED BY RAPID QUENCHING

MASAHIRO YOSHIMURA, TATSUO NOMA, YASUHIRO HANAUE, AND SHIGEYUKI SŌMIYA
Research Laboratory of Engineering Materials and Department of Materials
Science and Engineering, Tokyo Institute of Technology, 4259, Nagatsuta,
Midori, Yokohama 227 Japan.

ABSTRACT

Zirconia dispersed mullite ceramics were fabricated through hot-pressing at a pressure of 25 MPa from the amorphous material of 20wt%ZrO_2-80wt%($3Al_2O_3$·$2SiO_2$) obtained by rapid quenching of the melts. The densification was initiated at ∼950°C and accelerated by crystallization of mullite and t-ZrO_2 to yield almost pore free samples about 1050°C. This sample seemed to contain SiO_2-rich glassy phase in addition to Al_2O_3-rich mullite and t-ZrO_2. The further heat-treatments with/without pressing brought about the reaction between the SiO_2-rich glassy phase and the Al_2O_3-rich mullite phase to yield nearly stoichiometric mullite, and growth of the mullite grains and the t-ZrO_2 particles up to 1 μm in size at 1600°C. The TEM observation revealed a duplex microstructure of twinned m-ZrO_2 at the triple points and fine t-ZrO_2 in the center of the mullite grains. The fracture toughness of the specimen increased with hot-pressing temperatures up to 2.2 MPam$^{1/2}$ in the samples at 1600°C.

INTRODUCTION

Mullite ($3Al_2O_3$·$2SiO_2$) is one of the very important ceramics as a high temperature structural materials because of its good thermal and mechanical properties. As one of the ZrO_2 toughened ceramics, zirconia dispersed mullite ceramics has recently been studied by some investigators. Reaction sintering of $ZrSiO_4$ and Al_2O_3 was a popular method to prepare mullite-ZrO_2 ceramics [1-3]. The following reaction processes were proposed; "noncrystalline" (or transient liquid) phase was formed by the decomposition of $ZrSiO_4$ at first, ZrO_2 then crystallized and grew in this noncrystalline phase, and finally the noncrystalline phase crystallized into mullite phase. Zirconia dispersed mullite ceramics were also prepared by sintering of mixtures of ZrO_2 and some crystalline/noncrystalline mullite powders [4-6]. However, in the case of these inhomogeneous systems, the properties of the obtained samples were seriously depended upon the characteristics of the starting materials. The increase in the fracture toughness (K_{1c}) by the addition of ZrO_2 was observed in all those cases, however, various toughening mechanisms, transformation toughening or crack deflection toughening or grain boundary strengthening, were proposed in each cases.
We have already studied the formation and the crystallization of rapidly quenched amorphous materials in the system ZrO_2-SiO_2-Al_2O_3 [7-9]. The present study deals with ZrO_2 dispersed mullite ceramics through hot-pressing of the homogeneous amorphous material to study the microstructures, sintering behaviors, and mechanical properties of them.

EXPERIMENTAL

(1) Sample preperation

High purity ZrO_2 (99.9%), SiO_2 (guaranteed reagent), and Al_2O_3 (99.9%) powders were weighed to obtain the composition of $20wt\%ZrO_2-80wt\%(3Al_2O_3 \cdot 2SiO_2)$, and mixed throughly using an alumina ball mill for 20 h with 2wt% PVA aquaeous solution, then mixed again 4 h after drying. The mixtire was molded into a bar of ~4 mm diameter and 100-150 mm long under the isostatic pressure of 200 MPa. The molded samples were melted using a xenon arc-imaging furnace for floating-zone crystal growth (SC-5D, Nichiden Machinary Inc., Shiga, Japan). Molten droplets about 4 mm in size were rapidly quenched by passing through a steel twin- roller of ~1000 rpm to yield amorphous films typically with 60 μm thick and few 10 cm^2 area. The quenched amorphous films were ground into powder using an agate ball mill for 1 h with distilled water. The amorphous powder was hot-pressed under 25 MPa at 1050°, 1400° and 1600°C for 0.5 h using a graphite mold into a pellet of 10 mm diameter and 3-5 mm thick in an argon atmosphere. The heating rates during hot-pressing were ~2.2 °C/min to 1050°C, and 10 °C/min to 1400° and 1600°C. The sample hot-pressed at 1050°C were also heat-treated at 1200°, 1400°, and 1600°C for 1 h in air without pressing.

(2) Characterization methods

Crystalline phases of the samples were identified by X-ray powder diffraction. The microstructures of the samples were observed with a scanning (SEM) and transmission (TEM) electron microscopes after chemical etching or thermal etching polished samples for SEM observation, and after ion-beam thinning for TEM observation. A Micro-Vickers indentation method was applied to determine the fracture toughness (K_{1C}) of the samples at loads of 0.5, 1 and 2 kgf using following equation after Lawn et al.[10]

$$(K_{1C}/Ha^{1/2})(H/E)^{1/2}=0.028(c/a)^{-3/2} \tag{1}$$

and the (H/E) values were determined by a Knoop indentation method at the load of 0.5 kgf using following equation after Evans et al. [11]

$$b/a=0.142-0.45(H/E) \tag{2}$$

The densities of the hot-pressed samples were measured by a liquid substitution method using mercury.

RESULTS and DISCUSSION

(1) Densification

Figure 1 shows the densification curve of the sample during hot-pressing under 25 MPa at the heating rate of ~2.2 °C/min, and DTA curve of the amorphous material at the heating rate of 10 °C/min. The DTA curve showed a small endothermic peak at 919°C corresponding to the glass transition temperature, and an exothermic peak at 990°C corresponding to the crystalli- zation of the amorphous phase into mullite and t-ZrO_2 crystalline phases. The samples densified rapidly at ~950°C, around the glass transition and crystallization temperatures, indicating the crystallization of the amorphous phase accelerated the densification of the samples under the pressure of 25 MPa. Hot-pressing at 900°C for 0.5 h under 25 MPa resulted in little densification and crystallization as seen in Fig. 1. Hot-pressing at 950°C for 1 h under the same pressure brought about a shrinkage of ~20% and little crystallization of mullite. These results clearly indicate that the densification occurred in amorphous state by viscous flow as indicated in ZrO_2 containing cordierite glass ceramics [12], and was accelerated during crystallization. Recently, McPherson [13] also reported the preparation of the mullite-ZrO_2 composite by hot-pressing (with heating rate of 50 °C/min up to 1040°C under 21 MPa) of 13 to 19 μm sized glass powders containing 5wt%ZrO_2. The samples, however, had extensive porosity (31%). Our results, where the amorphous containing 20wt%ZrO_2 was fully densified by the hot-pressing at 1050°C with the heating rate of ~2.2 °C/min, suggest the heating rate is one of the critical parameters in the sintering. Further significant densification was not observed at higher temperatures than 1050°C up to 1600°C.

(2) Microstructures

Figure 2, the SEM photograph of the sample hot-pressed at 1050°C, shows a pore free microstructures consisted of 0.2-0.3 μm roundish mullite grains. The density of this sample is low (3.15 g/cm³) as compared to an expected value of 3.52 g/cm³ which had been calculated for 20wt%(t-ZrO_2)-80wt%(3:2mullite). This low density suggests that the samples hot-pressed at 1050°C contain a considerable amount of glassy phase in addition to mullite and t-ZrO_2 crystalline phases, although only mullite and t-ZrO_2 crystalline phases were identified by X-ray analysis in these samples. Figure 3 shows the relation between the lattice parameter (a_0) and the composition of the

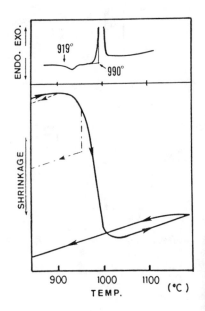

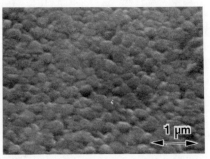

Fig. 2

Fig. 1
The densification curve during hot-pressing under 25 MPa, and DTA curve of the amorphous material. The sample densified rapidly about 950°C, around the glass transition and crystallization temperatures of the amorphous materials.
----- Hot-pressing at 900°C for 0.5 h under 25 MPa.
—·—·— Hot-pressing at 950°C for 1 h under 25 MPa.

SEM photograph of the sample hot-pressed at 1050°C. Although the microstructure of this sample was pore free, the density of this sample was much lower than the calculated one.

168

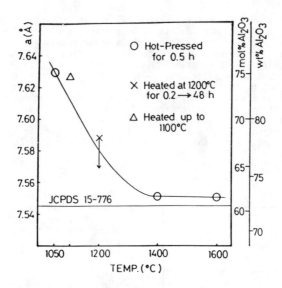

Fig. 3
Lattice parameter (a_0) of the mullite crystallized from rapidly quenched
amorphous materials as a function of temperature. The composition (right
axis) is derived from the data of lattice parameter (a_0) vs Al_2O_3 contents
of mullite in the system Al_2O_3-SiO_2 by Cameron (1977) [15].

mullite phase crystallized in the samples with various heating conditions.
The lattice parameter of the mullite phase in these samples hot-pressed at
1050°C was very large, a_0=7.629 Å in comparison with that for nearly stoichiometric mullite, a_0=7.546 A [16]. This result indicates that the mullite
crystallized in early stage from amorphous states has an Al_2O_3-rich composition [15]. The possible solubility of ZrO_2 in early crystallized mullite
has been discussed in another paper [17]. Such an Al_2O_3-rich phase would be
formed by a spinordal-like phase separation in the amorphous materials
during heating [9]. Figure 4, the TEM photograph of the sample hot-pressed
at 1050°C, shows fine (10-20 μm) t-ZrO_2 particles in the mullite grains,
especially along the grain boundaries. No significant region of glassy
phase was observed in this photograph. SiO_2-rich glassy phase would exist
in the mullite grains probably as small spheres due to a spinordal-like
phase separation during heating.
    Heat-treatment at higher temperatures of the samples hot-pressed at
1050°C brought about the reaction of the SiO_2-rich glassy phase and the
Al_2O_3-rich mullite. The densities of the heat-treated samples were increased
up to ~3.3 g/cm³ regardless to the treated temperatures above 1200°C because
of the disappearance of the SiO_2-rich glassy phase (Fig. 5). However, the
densities did not increase further because of the appearance of pores. Many
pores were observed in the samples heat-treated without pressing at 1400° and
1600°C (Fig. 6). As those pores could be removed by means of hot-pressing,
the densities of the samples increased up to 3.45 g/cm³ with pressing at
1600°C as seen in Fig. 5. This density is very close to the calculated one

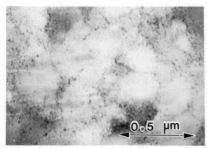

Fig. 4

TEM photograph of the sample hot-pressed at 1050°C. Fine t-ZrO$_2$ particles exists in the grains/along the grain boundaries of mullite. No signifcant glassy phase were observed in this sample.

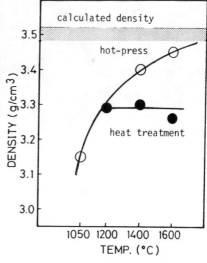

Fig. 5

The densities of the samples 20wt% ZrO$_2$-80wt%3Al$_2$O$_3$·2SiO$_2$ hot-pressed at various temperatures, or heat-treated after hot-pressed at 1050°C. Calculated density was derived from the densities of 3.18, 6.10 and 5.56 g/cm^3 for the stoichiometric mullite, t-ZrO$_2$ and m-ZrO$_2$, respectively.

Fig. 6

Polished surface of the sample heat-treated at 1600°C after hot-pressing at 1050°C.

for the sample containing m-ZrO$_2$ and 3Al$_2$O$_3$·2SiO$_2$ mullite. It means that the sample hot-pressed at 1600°C consists of fully crystallized phases of ZrO$_2$ and mullite with nearly stoichiometric composition. The lattice parameter of the mullite phase at this temperature (Fig. 3) supports this conclusion. We do not consider the solid solubility of ZrO$_2$ in mullite to be significant at ~1600°C, although Dinger et al. [14] recently suggested a high metastable solubility between mullite and ZrO$_2$ for the samples prepared at 1570°C by Moya et al. [4].

The microstructure of the sample hot-pressed at 1400°C, shown in Fig. 7, did not change significantly from the sample hot-pressed at 1050°C (Fig. 3) in size and shape of mullite grains. The TEM photograph, Fig. 8, however, showed apparent microstructural change at mullite grain boundaries. Fine t-ZrO$_2$ particles around mullite grain boundaries migrated to triple points to form larger grains. The grain growth of mullite became significant in the samples hot-pressed at 1600°C, where the grain size of mullite reached to 1 µm (Fig. 9). During the grain growth, the fine t-ZrO$_2$ particles were swept away by the motion of the mullite grain boundaries and gathered to make

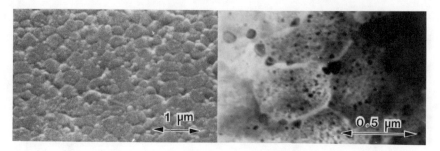

Fig. 7

SEM photograph of the sample hot-pressed at 1400°C. Almost unchanged from the sample hot-pressed at 1050°C but small ZrO_2 grains (white) were observed.

Fig. 8

TEM photograph of the sample hot-pressed at 1400°C. Large ZrO_2 grains were formed at mullite grain boundaries or triple points, due to the grain boundary diffusion.

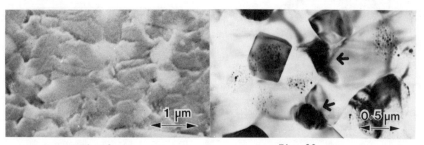

Fig. 9

SEM photograph of the sample hot-pressed at 1600°C. White grains of ZrO_2 are located in the triple points.

Fig. 10

TEM photograph of the sample hot-pressed at.1600°C. Twinned m-ZrO_2 (←) and fine t-ZrO_2 particles are seen.

larger grains at the triple points. But the fine t-ZrO_2 particles precipitated within the original mullite grains remained almost unchanged in the center of grown mullite grains. A duplex microstructure which consisted of large m-/t-ZrO_2 grains and mullite grains containing fine t-ZrO_2 particles has been observed in the sample hot-pressed at 1600°C as seen in Fig. 10. These large ZrO_2 grains of >0.5 μm at the triple points were transformable to m-ZrO_2, in fact, transformed m-ZrO_2 grains with twinning were seen in this TEM photograph.

(3) Mechanical properties

The K_{IC} values measured by micro-indentation method showed strong load dependences (Fig. 11). The values of 0.9 MPa·m$^{1/2}$ for hot-pressed sample at 1050°C, 1.2 MPa·m$^{1/2}$ for those at 1400°C, and 2.2 MPa·m$^{1/2}$ for those at 1600°C which were derived from sufficiently high load seem to be reasonable. The low K_{IC} value of 0.9 MPa·m$^{1/2}$ of the samples hot-pressed at 1050°C might be explained by the existance of SiO_2-rich glassy phase in these samples. The disappearance of the glassy phase brought about an increase of the K_{IC}

value to 1.2 MPa·m$^{1/2}$ in the sample hot-pressed at 1400°C. The significant increase of the K$_{IC}$ value in the sample hot-pressed at 1600°C could be explained by the transformation toughening of the large intergranular ZrO$_2$ grains. In many previous studies, other toughening mechanisms, for example, a grain boundary strengthening by a formation of metastable solid solution [4], have been suggested, but twinned m-ZrO$_2$ grains observed by TEM (Fig. 10) in the present study provide the evidence of the transformation toughening in the mullite-ZrO$_2$ ceramics. The comparison of the absolute value of K$_{IC}$, 2.2 MPa·m$^{1/2}$ in the present study, appears to be very difficult to the previous values (3-5 MPa·m$^{1/2}$) obtained for most mullite-ZrO$_2$ ceramics because the K$_{IC}$ value varies with the method and measuring conditions. The small sample size, 10 mmφ × 4 mm, prevented use of more reliable measurements for determining K$_{IC}$ values.

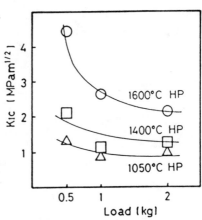

Fig. 11
The load dependences of the measured K$_{IC}$ values for the samples hot-pressed at various temperatures.

SUMMARY

(1) Zirconia dispersed mullite ceramics were obtained by hot-pressing of the rapidly quenched amorphous materials with the composition of 20wt%ZrO$_2$-80wt% (3:2mullite).

(2) The samples densified rapidly about 950°C, around the glass transition and and crystallization temperatures of the amorphous materials, probably due to viscous flow mechanism and enhanced by the crystallization.

(3) According to a spinodal-like phase separation, the samples hot-pressed at 1050°C seemed to contain SiO$_2$-rich glassy phase in addition to Al$_2$O$_3$-rich mullite and t-ZrO$_2$ crystalline phases. The glassy phase react with Al$_2$O$_3$-rich mullite phase leaving pores in the samples during further heat-treatments at higher temperatures without external pressing.

(4) The further heating with/without pressing yielded large ZrO$_2$ grains at the grain boundaries or triple points due to mass transportation along the mullite grain boundaries. The ZrO$_2$ grains grew up to 0.5-1 μm and were of transformable size in the sample hot-pressed at 1600°C.

(5) The fracture toughness of the specimens increased with hot-pressing temperatures due to the decrease of glassy phase, and also the additional tranformation of ZrO$_2$ up to 2.2 MPa·m$^{1/2}$ in the samples hot-pressed at 1600°C.

Acknowledgments

The authors thank to Professors E.Yasuda and M.Kato, and Miss M.Shibata for helpful experiments and discussion.

REFERENCE

1. J.S.Wallace, G.Petzow, N.Claussen, in Science and Technology of Zirconia II, Advances in Ceramics vol 12, edited by N.Claussen, M.Rühle, A.H.Heuer, (American Ceramic Society,Columbus, OH. 1984), p 436-442.

2. P.Boch, J.P.Giry, Materials Science and Engineering, 71, 39-48 (1985).

3. G.Orange, G.Fantozzi, F.Cambier, C.Leblud, M.R.Anseau, A.Leriche, J. Mater. Sci., 20, 2533-2540 (1985).

4. J.S.Moya, M.I.Osendi, J. Mater. Sci., 19, 2909-2914 (1984).

5. Qi-Ming Yuan, Jia-Qi Tan, Zheng-Guo Jin, J. Am. Ceram. Soc., 69, 265-67 (1986).

6. Qi-Ming Yuan, Jia-Qi Tan, Ji-Yao Shen, Xuan-Hui Zhu, Zheng-Fang Yang, J. Am. Ceram. Soc., 69, 268-69 (1986).

7. M.Yoshimura, M.Kaneko, S.Sōmiya, J. Mater. Sci. Lett., 4, 1082-84 (1985).

8. S.Sōmiya, M.Yoshimura, M.Kaneko, J. de Physique, C1, 47, 473-77 (1986).

9. M.Yoshimura, M.Kaneko, S.Sōmiya, Yogyo-Kyokai-Shi, 95, 202-208 (1987).

10. B.R.Lawn, A.G.Evans, D.B.Marshall, J. Am. Ceram. Soc., 63, 371 (1980).

11. D.B.Marshall, T.Noma, A.G.Evans, J. Am. Ceram. Soc., 65, c-175 (1982).

12. B.H.Mussler, M.W.Shafer, Am. Ceram. Soc. Bull. 64, 1459 (1985).

13. R.McPherson, J. Am. Ceram. Soc., 69, 297 (1986).

14. T.R.Dinger, K.M.Krishnan, G.Thomas, M.I.Osendi, J.S.Moya, Acta. Metall., 32, 1601 (1984).

15. W.E.Cameron, Ceram. Bull., 56, 1003 (1977).

16. JCPDS #15-776.

17. M.Yoshimura, M.Kaneko, Y.Hanaue and S.Sōmiya, Proceeding of the 3rd International Conference on the Science and Technology of Zirconia, Sept. 9-11, 1986, Tokyo, to be published.

PART IV

Mechanical Behavior of Reinforced Ceramic Composites

STRENGTH AND INTERFACIAL PROPERTIES OF CERAMIC COMPOSITES

D.B. MARSHALL
Rockwell International Science Center, 1049 Camino Dos Rios, Thousand Oaks,
CA 91360

ABSTRACT

    Results of recent micromechanics analyses of the reinforcing influence
of frictionally bonded fibers in ceramic composites are summarized. Direct
measurements of the fiber/matrix interface properties are also discussed.

INTRODUCTION

    Tensile strength and toughness of materials that are inherently brit-
tle can be dramatically improved by fiber reinforcement. Generally, this
requires a relatively low-toughness interface between the fibers and matrix
to permit debonding and sliding, and thereby allow fibers to bridge the
crack surfaces. Fracture mechanics models have been developed recently to
evaluate the influence of such bridging zones on the mechanisms of crack
growth. Solutions have been obtained for composites in which there is no
bonding at the interface, but sliding is resisted by friction. Novel
methods have also been devised for measuring the mechanical properties of
interfaces directly at individual fibers in weakly bonded composites. Some
results from the fracture mechanics analysis and interface measurements are
briefly summarized below. More complete descriptions are contained in
Refs. [1-6].

FRACTURE MECHANICS MODELING

    Fibers that bridge a crack cause a reduction of crack opening, and
thereby reduce the stress at the crack tip. The reduction of stress can be
evaluated by replacing the section of fiber between the crack surfaces by
closure tractions equal to the stress in the fiber and using a Green's
function [7] to calculate the stress intensity factor. The calculation
also requires evaluation of the crack opening displacements at every location in the crack, for the stress in each fiber is related, via the mechanics of the frictional sliding, to the crack opening. The displacements
were obtained by numerical solution of an integral equation derived by
Sneddon [8].

    In general, failure from a preexisting crack with a zone of bridging
fibers may occur by one of several sequences of events. Failure may initiate either by growth of the crack in the matrix or by failure of the
bridging fibers. Subsequently, each of these fracture processes can occur
either unstably or stably, leading to failure of the composite at constant
load or requiring further load increase to cause failure. The sequence
leading to failure must be evaluated in order to calculate the strength or
toughness of the composite. This entails calculation of the applied
stresses needed to cause both matrix crack growth and failure of the last
fiber in the bridging zone (i.e., the most highly stressed fiber) as a
function of both the total crack size and the length of the bridging zone.

    The results of such calculations for two extreme initial crack con-
figurations are summarized in Figs. 1 and 2. In Fig. 1, the applied
stresses necessary to cause matrix cracking and fiber failure, for an initially fully bridged crack, are plotted in normalized form (see Refs. [1

176

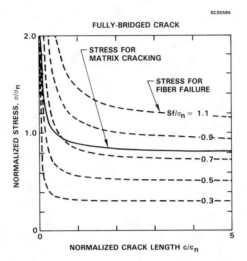

FULLY-BRIDGED CRACK

STRESS FOR
MATRIX CRACKING

STRESS FOR
FIBER FAILURE

$Sf/\sigma_n = 1.1$

0.9

0.7

0.5

0.3

NORMALIZED STRESS, σ/σ_n

NORMALIZED CRACK LENGTH c/c_n

Fig. 1 Applied stresses required to extend a fully bridged crack in the matrix and to fracture bridging fibers (after Ref. [1]).

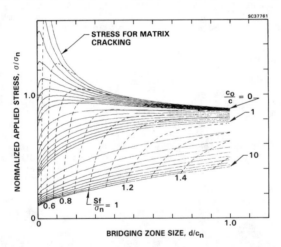

STRESS FOR MATRIX CRACKING

NORMALIZED APPLIED STRESS, σ/σ_n

$\frac{c_o}{c} = 0$

1

10

1.4

1.2

$\frac{Sf}{\sigma_n} = 1$

0.6   0.8

BRIDGING ZONE SIZE, d/c_n

Fig. 2 Applied stress needed to extend an initially unbridged crack in the matrix, plotted as a function of crack extension, d, for various initial crack lengths, c_o. Broken curves represent the condition at which catastrophic failure occurs, i.e., the bridging fibers fracture (after Ref. [1]).

and 3]) for various values of normalized fiber strength. For high-strength fibers ($Sf/\sigma_n > 0.8$, where S is the fiber strength, f the volume fraction of fibers, and σ_n a normalizing stress (see Eq. (1)), the stress for matrix cracking is lower than the stress for fiber failure, so that the crack grows in the matrix first. Moreover, the crack can extend indefinitely

without causing fiber fracture, and an increased applied stress is needed
to cause failure of the composite. This leads to a noncatastrophic mode of
failure involving multiple matrix cracking. This mechanism has been observed in glasses and glass-ceramics that are reinforced by carbon and SiC
fibers [9-12]. The steady-state matrix cracking stress in Fig. 1 is the
well-known solution of Aveston, Cooper and Kelly [9]:

$$\sigma_o = 0.8 \ \sigma_n = [\frac{6(1-\nu^2)K_o^2 \tau f^2 E_f E_c}{R(1-f)E_m^3}]1/3 \quad , \quad (1)$$

where K_o is the toughness of the unreinforced matrix, τ is the frictional
stress at the fiber/matrix interface, R is the fiber radius, ν is the
Poisson's ratio of the composite, and E_f, E_m and E_c are the elastic moduli
of the fibers, matrix and composite. For lower fiber strengths
(Sf/σ_n < 0.8), fiber failure occurs before matrix cracking, and is followed
by unstable matrix crack growth. Therefore, in this case, the composite
strength is determined by the stress for fiber failure.

A crack that initially has no bridging zone always grows stably in the
matrix with increasing applied stress. This stress is plotted as a function of crack extension for various initial crack lengths, c_o, in Fig. 2.
Stable growth continues until the stress in the bridging fibers builds up
to the critical value needed to break the fiber, whereupon the composite
fails catastrophically. This critical condition, for various normalized
fiber strengths, is indicated by the broken lines in Fig. 2. At the critical condition, the bridging effect of the fibers is equivalent to an increase in the fracture toughness given by

$$K_p/K_c = [1 + 4(Sf/\sigma_n)^3]^{1/2} -1 \quad , \quad (2)$$

where

$$(Sf/\sigma_n)^3 = (\frac{S^3 R}{12\tau K_o^2}) \ (\frac{f(1-f)E_m^3}{(1-\nu^2)E_f E_c^2}) \quad . \quad (3)$$

PROPERTIES OF THE INTERFACE

It is clear from Eqs. (1) and (3) that the magnitude of the interfacial frictional stress plays a key role in determining the strengthening
and toughening, as well as the failure mechanism of the composite. A
method for measuring the frictional stress at individual fibers is illustrated schematically in Fig. 3. The technique involves pushing the end of
a fiber with a sharp indenter and measuring the resultant displacement of
the fiber beneath the surface of the matrix. Analysis of the mechanics of
fiber sliding under this condition allows the frictional stress to be obtained from measurements of the force applied to the fiber and the
displacement.

A standard Vickers hardness testing instrument can be used to obtain
force and displacement measurements at the peak load condition [4], thereby
providing an average value of the frictional stress over the area of interface that undergoes sliding. However, more information can be obtained
using an instrument that allows continuous measurement of force and displacement during loading (and unloading) [6]. Results obtained from a SiC/

178

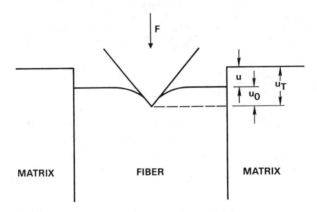

Fig. 3   Schematic diagram of indentation experiment to measure interfacial sliding resistance (after Ref. [6]).

glass-ceramic composite* using a nano indenter instrument** are shown in Fig. 4. The curves represent predictions of the force-displacement relation for purely frictional sliding with constant frictional stress during loading, unloading, and reloading. The data follow the prediction very closely during initial loading, indicating that the frictional stress is uniform along the interface (each force increment causes the area of interface over which fiber/matrix sliding occurs to increase). However, during unloading and subsequent reloading, when reverse sliding occurs, the results indicate that the frictional stress decreases.

The results in Fig. 4 also indicate that sliding at the interface in this composite does not require prior debonding. Analysis of combined debonding and frictional sliding during the initial loading indicates that the force-displacement relation becomes [6]

$$u = F^2/4\pi^2 R^3 \tau E_f - 2\Gamma/\tau \quad , \tag{4}$$

where Γ is the fracture surface energy associated with mode II debonding. Fitting Eq. (4) to the data in Fig. 4 gives $\tau = 2.9$ MPa and $\Gamma \leq 0.4$ J/m². This upper bound for the value of Γ, obtained by taking into account maximum measurement errors for the data in Fig. 4, is in the range of energies associated with Van der Waals bonds.

ACKNOWLEDGEMENTS

The research summarized here was done in collaboration with B.N. Cox, Rockwell International Science Center, W.C. Oliver, Oak Ridge National Laboratory, and A.G. Evans, University of California at Santa Barbara. Funding was provided by the U.S. Office of Naval Research, Contract No. N00014-85-C-0416.

---

* United Technologies Research Center, East Hartford, CT
** Microscience Inc., Braintree, MA

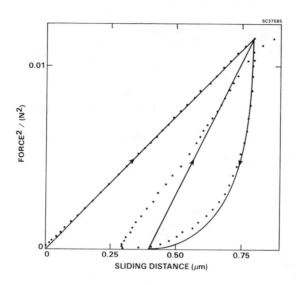

Fig. 4 Forces and displacements measured during indentation (loading, unloading, reloading) of fiber as in Fig. 3. Solid curves are theoretical predictions for sliding opposed by constant frictional stress (after Ref. [6]).

REFERENCES

1. D.B. Marshall and B.N. Cox, "Tensile Properties of Brittle Matrix Composites: Influence of Fiber Strength," to be published in Acta Met.
2. D.B. Marshall, B.N. Cox and A.G. Evans, Acta Met. 33 (11), 2013-21 (1985).
3. D.B. Marshall and A.G. Evans, in Fracture Mechanics of Ceramics 7, Ed., R.C. Bradt, A.G. Evans, D.P.H. Hasselman and F.F. Lange (Plenum), 1986, pp. 1-15.
4. D.B. Marshall, J. Am. Ceram. Soc. 67 (12), C259-60 (1984).
5. D.B. Marshall, in Ceramic Microstructures '86: Role of Interfaces, eds., J.A. Pask and A.G. Evans (Plenum, in press).
6. D.B. Marshall and W.C. Oliver, J. Am. Ceram. Soc., in press.
7. G.C. Sih, Handbook of Stress Intensity Factors, Lehigh University Press, Bethlehem, PA (1973).
8. I.N. Sneddon and M. Lowengrub, Crack Problems in the Classical Theory of Elasticity, Wiley, NY (1969).
9. J. Aveston, G.A. Cooper and A. Kelley, in Properties of Fiber Composites, Conf. Proc. Nat. Physical Lab. IPC Science and Technology Pres. Ltd., Surrey, England, 1971, pp. 15-26.
10. D.B. Marshall and A.G. Evans, J. Am. Ceram. Soc. 68 (5), 225-31 (1985).
11. J.J. Brennan and K.M. Prewo, J. Mater. Sci. 17 (8), 2371-83 (1982).
12. R.A.J. Sambell, A. Briggs, D.C. Phillips and D.H. Bowen, J. Mater. Sci. 7 (6), 676-81 (1972).

THE STRENGTH OF FIBRES IN ALL-CERAMIC COMPOSITES

KEVIN KENDALL, N. McN. ALFORD AND J.D. BIRCHALL
ICI New Science Group, P.O. Box 11, The Heath Runcorn, Cheshire, UK.

ABSTRACT

When considering the strength of a fibre reinforced ceramic composite, it is often assumed that the fibres retain their full strength of several GPa after cracking of the weaker matrix. The strength of the composite after matrix cracking is then calculated by the rule of mixtures as the product of fibre volume fraction and fibre strength. This paper demonstrates that such a calculation is not consistent with the principles of fracture mechanics for an isolated fibre embedded in an elastic matrix of the same elastic modulus, because the strength of the fibre is much reduced by the stress concentration arising from the matrix crack. Experimental measurements of the strength of a glass fibre embedded in a brittle matrix support the theory. The case of a fibre in a matrix of different elastic modulus is also considered, together with the problem of cracking along the fibre-matrix interface.

INTRODUCTION

Since the earliest studies of fibrous reinforced materials, it has been presumed that the strength of a composite can be predicted from a simple mixtures rule [1]. In the case of a uniaxial, continuous fibre composite containing a large volume fraction of fibres, the weak matrix fails first under tension and so transfers all its load onto the fibres. Thus the ultimate strength σ_{cu} of the composite in tension is given by:-

$$\sigma_{cu} = \sigma_{fu} V_f \qquad \cdots (1)$$

where σ_{fu} is the strength of each fibre and V_f the volume fraction of fibres in the composite.

There have been many experimental tests of this relation for brittle fibres embedded in a soft metallic matrix [2]. Except at low levels of fibre, the measured strengths (Fig. 1) give a reasonable fit to equation 1 [3].

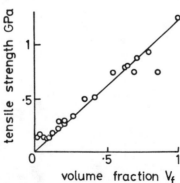

Fig. 1   Tensile strength results [3] for copper matrix reinforced with continuous tungsten wires at various volume fractions.

182

Although modern text-books on composite materials [4] support equation 1
there is a problem that the above theory of strength is not consistent with
the theory and practice of fracture mechanics. According to the energy
principle first enunciated by Griffith [5], failure of a brittle material
is determined by the stress at which a crack propagates through it, and
this stress depends not on an intrinsic strength parameter but on the
length of the crack, the fracture energy of the material and its elastic
properties. On this basis, we contend that reinforcing fibres do not have
an intrinsic strength but must be influenced by the matrix except under
special circumstances. For example, a glass matrix reinforced with 70% by
volume of glass fibres of strength 1.5 GPa should have given a strength of
1.05 GPa according to equation (1). Experimentally, the strength was 0.25
GPa. What is wrong with the mixtures rule for strength of composites?

MIXTURES RULE

The mixtures rule accounts for the elastic deflections in a composite
under stress. Figure 2a shows such a composite with a long uniaxial fibre
under a stress σ_f, strain ε_f, of modulus E_f and volume fraction V_f. For
the matrix the subscript m applies and for the composite, the subscript c.
For a well-bonded composite, the matrix and fibres are constrained to move
together so that:-

$$\varepsilon_c = \varepsilon_f = \varepsilon_m \qquad \dots (2)$$

Adding the loads on fibre and matrix gives the total load on the
composite.

$$\begin{aligned}
\sigma_c &= \sigma_f V_f + \sigma_m (1-V_f) \\
&= E_f \varepsilon_f V_f + E_m \varepsilon_m (1-V_f) \\
&= \varepsilon_c (E_f V_f + E_m (1-V_f)) \qquad \dots (3)
\end{aligned}$$

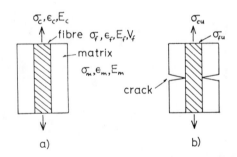

Fig 2 a) Model of a brittle matrix, modulus E_m under strain ε_m and σ_m
            reinforced with unidirectional fibres of volume fraction V_f,
            modulus E_f under a stress σ_f and strain ε_f, to give a
            composite structure of modulus E_c and strain ε_c under a
            stress σ_c.
        b) Model for a ceramic composite whose brittle matrix has
            cracked under tension to transfer all the load onto the
            brittle fibres of strength σ_{fu}.

The composite modulus $E_c = \sigma_c / \varepsilon_c$

Therefore $E_c = E_f V_f + E_m (1-V_f)$ .... (4)

Thus the mixtures rule applies to the elastic properties of the composite.

Now imagine that the matrix cracks first on stretching the composite (Fig. 2b) so that all the load across the crack is transferred onto the fibres which are assumed to retain their original strength σ_{fu}. Then the matrix supports no load across the crack and the ultimate stress σ_{cu} supportable by the composite from equation 3 is given by equation 1. Other more complicated expressions for composite strength have also been derived from the mixtures rule [4]. These are all based on the notion that the strength of a composite can somehow be calculated, like elastic deflections, by summing the strengths of the components.

PROBLEM

The problem is that equation (1) is not consistent with the teaching of fracture mechanics. Consider an isolated fibre under tension (Fig. 3a). Viewed from the standpoint of fracture mechanics, this fibre contains a small flaw of diameter 2c (much smaller than the fibre diameter d) and failure of the fibre is due to propagation of this flaw under stress. For a disc shaped flaw the propagation stress is given by [6,7].

$$\sigma = \left[\frac{\pi E_f R_f}{4c(1-\nu^2)}\right]^{\frac{1}{2}} = K_{1c} \left[\frac{\pi}{4c(1-\nu^2)}\right]^{\frac{1}{2}} \qquad .... (5)$$

where R_f is the fracture energy of the fibre, ν the Poisson's ratio and K_{1c} its fracture toughness. For example, a 30 µm diameter alumina fibre of K_{1c} = 4.5 MPa $\sqrt{m}$ containing a disc shaped flaw of radius 2.4 µm exhibits a strength of 2700 MPa.

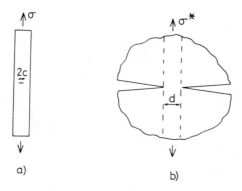

a)

b)

FIG 3 a) An isolated brittle fibre under tension fails when a penny shaped crack of radius c propagates.
     b) A well-bonded fibre embedded in a matrix of equal modulus must be influenced by the matrix crack, and fails at a lower stress σ^*.

184

Next consider a similar fibre embedded in a cracked brittle matrix (Fig. 3b). This fibre is now acted on by two cracks:- the intrinsic flaw in the fibre and the crack attacking the fibre surface from the matrix. The stress σ^* required to propagate the matrix crack into the fibre can be calculated when the matrix and fibre have equal modulus [8,9], if the flaw in the fibre is ignored.

$$\sigma^* = \left[\frac{8\ E_f R_f}{\pi d(1-\nu^2)} \right]^{\frac{1}{2}}$$

.... (6)

where d is the fibre diameter. For the 30 μm diameter alumina fibre, σ^* would be 1317 MPa, only half the strength of the free fibre given by equation (5). It is clear from equations (5) and (6) that large diameter fibres must appear weak in a cracked brittle matrix of the same elastic modulus. Therefore, equation (1) giving the strength of a composite in terms of fibre strength is not generally correct. Only when the fibre diameter is reduced to be comparable with the intrinsic flaw size c (i.e. d ≈ 32c/π²) does the fibre attain its full strength in the cracked matrix.

EXPERIMENTAL

The purpose of the experiments was to check equation (6) for the strength of a single fibre encased in a cracked matrix of the same elastic modulus.

In the first test, polymethylmethacrylate was used to simulate a brittle fibre, perfectly bonded to a brittle matrix of equal elastic modulus. Polymethylmethacrylate rods were turned on a lathe to produce a deep V shaped notch, as in Fig. 3b. The diameter d of the remaining ligament was measured and $d^{-\frac{1}{2}}$ was plotted against the failure stress determined by tensile testing. Figure 4 shows the results for several values of d, and also gives the ultimate strength of unnotched polymethylmethacrylate rods, for comparison with the predictions of equation (6) taking $\nu = 0.3$ and $K_{1c} = 1.2$ MPa √m, measured in single edge notched tests [10].

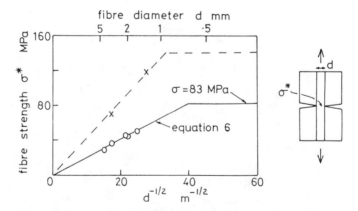

Fig 4   Comparison of theory (-) with experimental results (0) for fracture of notched polymethlmethacrylate rods. Also shown are results (x) for fracture of a glass rod embedded in MDF cement of 54 GPa elastic modulus.

The results fitted equation (6) closely for large diameters d, but as
d was decreased, the strength approached that of the unnotched rod,
becoming constant at a value of 83 MPa, which corresponded to a flaw size
of c = .18 mm from equation 5. It follows that a brittle fibre, well-bonded
to a cracked matrix of the same elastic modulus, will be weakened according
to equation (6).

This result is relevant, for example, to composites of silicon carbide
fibres in a silicon carbide matrix [11] where the components have the same
elastic modulus.

INFLUENCE OF MODULUS MISMATCH

In practice, it is more common to find a modulus mismatch between
fibre and matrix, for example when carbon fibres reinforce a glass matrix
[12]. In this case, the matrix crack is trying to penetrate a higher
modulus fibre, and although there is no precise theory to predict the
fracture condition in this case, it is known that the stress will be higher
than that given by equation (6). Similarly, if the fibre has a lower
modulus than the matrix, a lower stress will be needed to crack the fibre
[13,14].

To estimate the influence of elastic modulus mismatch, borosilicate
glass rods were embedded in a matrix of MDF cement [15,16] of elastic
modulus 54 GPa and the tensile strength was measured. The rods were found
to crack at the point where they emerged from the MDF cement matrix and
their strength was found to depend on $(diameter)^{-\frac{1}{2}}$ as expected from
equation (6). The failure stress was much less than the 140 MPa strength of
the same rods tested in air. Figure 4 shows some results.

By varying the elastic modulus of the MDF cement systematically, using
alumina particles to stiffen the matrix, it was possible to repeat this
experiment for several ratios of fibre modulus/matrix modulus (E_f/E_m) at a
fibre diameter of 3.2 mm. The strength of the fibre was observed to rise
rapidly, nearly in proportion to $(E_f/E_m)^2$, as shown in Fig. 5. There was
no sign of interfacial debonding of fibres in these experiments. Such
interfacial fracture would invalidate the foregoing arguments [14].

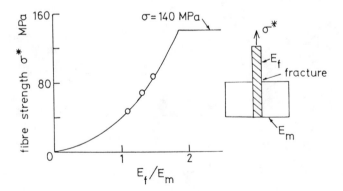

FIG 5   Observed increase in fibre strengths when embedded in a cracked
matrix of lower elastic modulus.

CRACKING AT FIBRE MATRIX INTERFACE

It is known that cracking can occur at the fibre-matrix interface before fibre fracture. For the case of a fibre in a matrix of the same elastic modulus, this cracking is similar to that caused by indentation of a rigid cylindrical punch into an elastic material, a problem considered by Mouginot and Maugis [17]. They showed that there were two possibilities depending on the ratio $2c/d$ of interface defect size to fibre diameter (Fig 6).

For very small flaws ($2c/d < .01$) the criterion for interface cracking is

$$\sigma = \left[2E\, R_{int}/\pi c(1-\nu^2)(1-2\nu)^2\right]^{1/2} \tag{7}$$

where R_{int} is the interface fracture energy. If this stress σ is less than that required to crack across the fibre, given by equation (6), then interface fracture will be initiated first. Comparing equations (6) and (7), it is clear that interface fracture occurs when

$$R_f/R_{int} > d/4c(1-2\nu)^2 \tag{8}$$

Thus, for small interface flaws, the interface adhesion energy is small and dependent on flaw size if debonding is to occur (Fig 6).

However, for larger interface flaws ($2c/d > .01$), the condition for interface cracking changes to

$$\sigma = \left[\pi\, ER_{int}/4d(1-\nu^2)\phi\right]^{1/2} \tag{9}$$

where ϕ is a tabulated number which depends particularly on Poisson's ratio [17]. For $\nu = .22$, ϕ is 10^{-3}. Comparing equation (9) with equation (6), it is evident that interface fracture will be preferred over fibre fracture when

$$R_f/R_{int} > \pi^2/32\, \phi \tag{10}$$

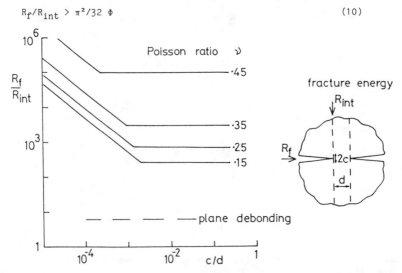

Fig. 6   Variation of critical fracture energy ratio R_f/R_{int} for interface cracking with interface flaw size ratio c/d and Poisson's ratio ν.

Therefore, large interface flaws cause interface fracture at higher
interface energies, and the value is independent of flaw size and fibre
diameter but strongly varying with Poisson's ratio as indicated in figure
6. Interestingly, the fibre debonding requires lower interface fracture
energy than simple plane delamination where the propagation criterion is
[18]

$$R_f/R_{int} > 2\pi^2(1-\nu^2)$$

CONCLUSION

A brittle ceramic fibre embedded in a cracked ceramic matrix of equal
elastic modulus must be weaker than the same fibre tested freely unless the
fibre diameter is comparable with its intrinsic flaw size. This result
stems directly from the theory of fracture mechanics and has been verified
experimentally. When the matrix has a lower modulus, the fibre strength
increases substantially as shown by model experiments using borosilicate
rods embedded in MDF cement of controlled elastic properties.

Therefore, the rule of mixtures cannot generally be applied to
calculate the strength of all-ceramic composites. However, when the fibres
are sufficiently thin, when the matrix has a sufficiently low elastic
modulus, or when cracking occurs along the fibre-matrix interface, the
fibres can reveal their full strength.

REFERENCES

1. D.L. McDanels, R.W. Jech and J.W. Weeton, Metal Prog. 78 (6), 118
    (1960).
2. A. Kelly and G.J. Davies, Metall. Rev. 10, 1 (1965).
3. A. Kelly, Strong Solids, (Clarendon Press, Oxford 1966), p.140
4. D. Hull, An introduction to composite materials (Cambridge University
    Press, Cambridge 1981) p.127.
5. A.A. Griffith, Phil. Trans. R. Soc. Lond. A221, 163 (1920).
6. R.A. Sack, Proc. Phys. Soc. 58, 729 (1946).
7. I.N. Sneddon, Proc. R. Soc. Lond., A187, 229 (1946).
8. J. Boussinesq, Application des Potentiels à l'etude de l' équilibre et
    du mouvement des solids élastiques (Gauthier-Villars, Paris 1885).
9. J.P. Benthem and W.T. Koiter, Mechanics of Fracture Vol. 1 edited by
    G.C. Sih (Noordhof, Leyden 1973) p.151.
10. K. Kendall, J. Mater. Sci. Letters, 2, 115 (1983).
11. P L Lamicq, E A Bernhart, M M Dauchier and J G Mace, Am. Ceram. Soc.
    Bull. 65, 336 (1986).
12. R.A. Sambell, D. Bowen and D.C. Phillips, J. Mater. Sci. 7, 663
    (1972).
13. K. Kendall, MRS Symposium Proceedings, Vol 40 (eds. Giess Tu and
    Uhlmann) (MRS Pittsburgh 1985) 167.
14. K. Kendall, Proc. R. Soc. Lond. A341, 409 (1975).
15. J.D. Birchall, A.J. Howard and K. Kendall, Nature 289, 388 (1981).
16. K. Kendall, A.J. Howard and J.D. Birchall, Phil. Trans. R. Soc. Lond.
    A310 139 (1983).
17. R. Mouginot and D. Maugis, J. Mater. Sci. 20, 4354 (1985).
18. K. Kendall Proc. R. Soc. Lond., A344, 287 (1975).

WEIBULL MODULUS OF TOUGHENED CERAMICS

KEVIN KENDALL, N.McN.ALFORD AND J.D.BIRCHALL
ICI New Science Group, P.O.Box 11 The Heath, Runcorn, Cheshire, UK

ABSTRACT

The influence of toughness on the Weibull modulus of ceramics is discussed. It is shown that an increase in toughness does not give a higher Weibull modulus if the ceramic is truly brittle, following the Griffith criterion of fracture. However, R curve behaviour, such as that shown by tetragonal zirconia, by Dugdale material, or by fibrous ceramic composites, leads to an improved Weibull modulus. Theoretical argument and experimental results support these conclusions.

INTRODUCTION

The principal reason for toughening a ceramic is to inhibit the growth of long cracks which may be present in the material owing to poor processing or subsequent damage. Such toughening is revealed by the increase in applied stress necessary to break a notched sample of the ceramic. For example, adding ceramic whiskers to alumina increases the value of K_{1C} from 4.5 to 8.7 MPa $\sqrt{m}$ [1]. Putting aligned silicon carbide filaments into a silicon carbide matrix raises K_{1C} from 5 to 27 MPa $\sqrt{m}$ [2]. Transformation toughening of zirconia can lift K_{1C} from 4 to 9 MPa $\sqrt{m}$ [3].
A second benefit of some toughened ceramics is their increased reliability. When the long cracks are inhibited by the reinforcing mechanism more than the short cracks, the distribution of flaws in the ceramic is effectively narrowed and the Weibull modulus of strength values (ie reliability) is raised. For example, transformation toughened zirconia can display a Weibull modulus of 20 (Fig 1) but this drops to a more ordinary value of 10 when the toughening is lost above the transformation temperature [4].

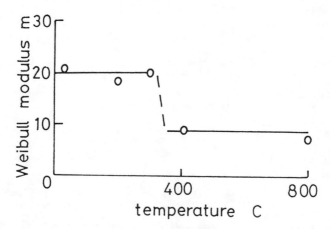

Fig.1. Fall in Weibull modulus of partially stabilised zirconia when toughening is lost above the transformation temperature.

This influence of toughening on reliability is not well understood. Recent work has shown that ceramics do not necessarily become more reliable as they are made tougher [5]. The purpose of this paper is to consider when Weibull modulus should be raised by the toughening process. First, the behaviour of truly brittle or "Griffith" [6] ceramics is treated. Then the influence of transformation toughening caused by residual strain energy is discussed. This leads on to the case of plastic toughening as exemplified by a "Dugdale" [7] material, and finally on to the problem of fibre reinforced ceramics.

GRIFFITH BEHAVIOUR

Many ceramics are almost perfectly brittle. That is, they obey the Griffith cracking criterion for tensile fracture of a centre-through-cracked thin plate giving a strength

$$\sigma = (ER/\pi c)^{\frac{1}{2}} = K_{1C}/\sqrt{\pi c} \qquad (1)$$

where E is the Young's modulus, R the fracture surface energy, K_{1C} the toughness of the material and 2c is the length of the crack. For edge cracks equation (1) must be modified to

$$\sigma = K_{1C}/1.122 \sqrt{\pi c} \qquad (2)$$

where c is the length of the edge crack. It is clear, therefore that edge flaws reduce strength by a factor 1.59 when compared with flaws of the same length in the body of the material. For penny shaped flaws totally embedded in a large block of brittle sample, the Sack [8] equation gives the relation between flaw diameter 2c and strength σ where

$$\sigma = K_{1C} (\pi/4c (1-\nu^2))^{\frac{1}{2}} \qquad (3)$$

This means that an enclosed flaw gives an increased strength by a factor 1.62 compared with a through crack, when $\nu = .25$, almost exactly off-setting the factor 1.59 edge crack weakening. Thus, equation (1) gives a good description of the strength resulting from enclosed edge flaws of diameter 2c [9].

When several ceramic samples are tested, the strength values may be plotted on a Weibull [10] distribution given by

$$\ln \ln \frac{1}{1-P_f} = m \ln \sigma - m \ln \sigma_0 \qquad (4)$$

where P_f is the probability of failure (P_f being defined as $n/(q+1)$, n being the number of samples with strength σ or less and q the total number of samples tested), m is the Weibull modulus and σ_0 is a constant. Effectively, the Weibull modulus m is determined by the ratio of the highest strength σ_{max} to the lowest strength σ_{min} in the batch tested by the equation

$$m = \frac{\ln \ln (q+1) - \ln \ln ((q+1)/q)}{\ln (\sigma_{max}/\sigma_{min})} \tag{5}$$

where q is the total number of samples tested.

Consider the application of these ideas to a ceramic with Griffith behaviour (Fig 2) of toughness K_{1C}. The range of strength results is shown by the arrow along the strength axis, with a maximum result of σ_{max} and a minimum of σ_{min}. From equation (1), these results correspond to edge flaw radii of c_{min} and c_{max} respectively, as shown by the arrow on the c axis.

Now imagine that the toughness of the material is raised to K_{1C2}, while the material still follows the Griffith equation (1). If the flaw sizes remain the same, then the measured strength σ_{max} and σ_{min} for the same series of strength tests will both be raised by the ratio K_{1C2}/K_{1C}. Therefore, the ratio of $\sigma_{max}/\sigma_{min}$ will be the same as before and it is obvious from equation (5) that the Weibull modulus will not change.

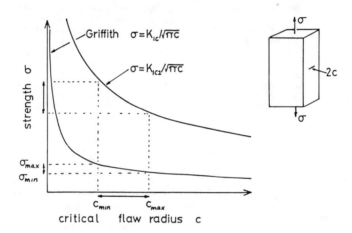

Fig.2. Connection between strength σ and flaw radius c for Griffith behaviour of a ceramic. A range of flaw radii from c_{min} to c_{max} gives a range of strengths from σ_{max} to σ_{min}. Higher toughness does not change $\sigma_{max}/\sigma_{min}$.

This theory was verified experimentally by measuring the bend strengths of titanium dioxide powder compacts before and after sintering [11]. The TiO_2 powder (RCR2, Tioxide) was mixed into an aqueous solution of polyvinylalcohol/acetate (KH17S, Nippon Gohsei) and dried to produce a sheet containing 61% by volume of ceramic. After cutting into strips, the polymer was burnt out by heating at 1°/min to 450°C and strength tested to give the results shown in Fig 3 (left hand curve). Strips from the same batch were fired at 1300°C for 1 hour to give a dense ceramic whose bend strengths are plotted in Fig 3 (right hand curve). Despite the increase in K_{1C} from .036 to 2.0 MPa $\sqrt{m}$ on sintering, the Weibull modulus did not change substantially from m = 7 [5]. Therefore, toughening does not increase Weibull modulus of a "Griffith" material.

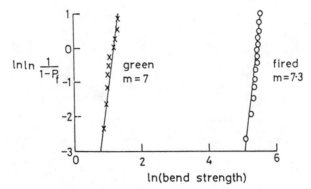

Fig.3. Strength results plotted for comparison with the Weibull equation, showing that green ceramic of K_{1C}=.036 MPa √m gives essentially the same Weibull modulus as fired ceramic of K_{1C} = 2.0 MPa √m.

TRANSFORMATION TOUGHENING

When zirconia is transformation toughened, it ceases to follow the Griffith description and consequently, there is a problem in defining the strength and toughness of such materials. In the absence of an exact theory to describe failure of toughened materials, we adopt the pragmatic engineering approach of defining strength σ as the stress at which rapid fracture occurs, and crack length c as the original length of notch in the sample. Fracture experiments on zirconia then give results like those in the upper R curve of Fig. 4a. Although the results do not follow the Griffith line (Fig 4a, lower curve) the nominal toughness K_{1C} may be defined at any point on the R curve as σ √πc (for a centre-through-notched thin plate). Toughness therefore seems to increase with crack length (Fig. 4b). In reality, of course, the orginal crack grows in a stable fashion before rapid fracture occurs [17] but this is of little interest to the engineer, as in the precise nature of K at the crack tip.

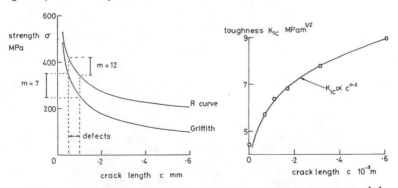

Fig.4. a) R curve behaviour of transformation toughened zirconia [3] leads to an increase in Weibull modulus compared to Griffith behaviour for the same spread of defects.

b) Increase in toughness with crack length is reasonably described by a power law.

In Fig 4a, a spread of edge defect radii from .05 mm to 0.1 mm has been considered. On the Griffith basis (equation (1)) this corresponds to strength results from 248 to 351 MPa, but on the shallower R-curve the strengths are higher from 344 to 423 MPa. Thus, if the Weibull modulus of brittle zirconia were 7, then that for R-curve zirconia would be 11.76, an increase by a factor of 1.67.

This case is a simple one because the increase in K_{1C} with crack length is reasonably described by a power law over much of the crack length range (Fig 4b), corresponding to the equation

$$K_{1C} = 39c^{0.2} = Ac^n \tag{6}$$

where K_{1C} is in MPa $m^{\frac{1}{2}}$ and C is in metres. Thus, the strength σ_R of R-curve material is

$$\sigma_R = Ac^n/(\pi c)^{\frac{1}{2}} = A/\sqrt{\pi}\ c^{(.5-n)} \tag{7}$$

Hence, from equation (5) is is obvious that the Weibull modulus m_R for the R curve material must be larger than that m_G for the Griffith material by the factor

$$m_R/m_G = 0.5/(.5-n) \tag{8}$$

which corresponds to 1.67 for the material shown in Fig 4.

PLASTIC TOUGHENING

Plastic deformation around a crack must also lead to a deviation from the Griffith theory because the applied load is now doing irreversible work on the bulk of the material in addition to the elastic work assumed in the original Griffith formulation.

For the Dugdale model of a stretched sheet at stress σ containing a Griffith crack of length 2c tied by plastic stress σ_y to confine the opening to a length 2a (the Dugdale crack length), shown in Fig 5a, the following relationship holds

$$a/c = \cos(\pi\sigma/2\sigma_y) \tag{9}$$

For a plastic material, σ_y is the yield stress.

This behaviour is shown in Fig 5b, which shows the stress required to propagate notches cut in a titanium dioxide green body containing polymer binder. This sample was made as before by mixing TiO_2 powder into aqueous polyvinyl alcohol solution, pressing the plastic mix to a sheet and drying. On cracking these green strips, it was found that the results gave a better fit to equation (9) than the Griffith curve. The yield stress σ_y was taken to be 47 MPa.

In a series of ten strength tests on unnotched material, the mean strength was 44 MPa with a Weibull modulus of 20. On the Dugdale curve this corresponds to flaw sizes between 50 and 150 μm. On the Griffith curve, such flaws would give strengths between 82 and 51 MPa, giving a Weibull modulus of 6.5, a much lower value. When the TiO_2 green body was burned-out to remove the polymer giving Griffith behaviour, the Weibull modulus was 7 and after subsequent sintering was 7.3, closely corresponding to the theoretical prediction of 6.5. It is clear from these experiments that plasticity can substantially increase Weibull modulus, in accord with theoretical prediction using the Dugdale model.

194

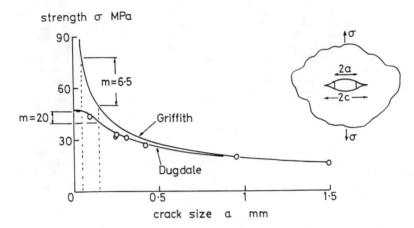

strength σ MPa

Fig 5 a) Dugdale model of a tied crack
       b) Results showing Dugdale behaviour and enhanced Weibull
          modulus

From Fig 5b, it is obvious that the reliability of the plastic
material at large flaw sizes will be similar to the brittle material
because the Dugdale and Griffith curves converge. However, for small flaws
the reliability of the plastic material must increase because the stress
required to crack the plastic material approaches the yield stress, thus
limiting the range of failure stresses for the same flaw size distribution.
This increase in Weibull modulus as flaw size is diminished can be
calculated from the Dugdale model for a typical ceramic such as alumina,
assuming that K_{1C} = 4 MPa √m and σ_y = 7000 MPa, using the more convenient
description of the Dugdale crack [12].

$$\sigma = 2\sigma_y \ (\cos^{-1} \ \exp - \ [\pi E\delta/8\sigma_y a])/\pi \qquad (10)$$

where δ is the crack opening displacement ($\approx K_{1C}^2/E\sigma_y$).

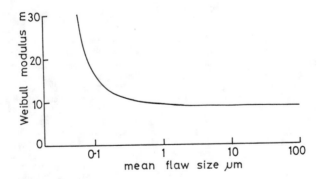

Fig. 6. Expected increase in Weibull modulus as flaw size
        is decreased in alumina, keeping the flaw size
        distribution constant, following the Dugdale model.

Imagine that a Weibull distribution of critical flaws of size a=50 to a=100 μm exists in a series of 9 test samples, giving strengths ranging from 226 MPa to 319 MPa and a Weibull modulus of 8.90 (Fig 6). Now consider shrinking the flaws by a factor 10 so that a=5 μm for the smallest flaw and a=10 μm for the largest. The samples still behave substantially in a Griffith fashion but the strengths range from 712 MPa to 1005 MPa and the Weibull modulus has risen only slightly to 8.95. However when the flaws are further reduced in size such that the fracture stress approaches the yield stress, the Weibull modulus of strength values must increase as the material follows the Dugdale rather than the Griffith curve. This increase becomes noticeable for flaws less than 0.5 μm. For 0.1 μm flaws the Weibull modulus doubles (Fig 6).

FIBRE TOUGHENED CERAMICS

There is no doubt that the addition of fibre reinforcement to a ceramic material can improve its Weibull modulus. Values of 24 have been achieved with silicon carbide whiskers in a silicon nitride matrix [13]. The problem with fibre reinforcement is that the Weibull modulus sometimes drops. The effect of reinforcement on Weibull modulus is not understood.

Although it has been suggested that "High toughness leads to flaw tolerance and high Weibull modulus" [14], extra toughness alone cannot account for the reliability increase, as demonstrated in Figure 2. The real reason for the rise in Weibull modulus is R-curve behaviour which is often observed in fibre reinforced ceramics.

Experimental results for the increase of toughness with crack length are shown in Fig 7a for silicon carbide fibre reinforced silicon carbide [2]. Toughness starts at a level somewhat above that of the matrix but then rises substantially as damage spreads around the crack tip, eventually reaching a limit at a crack length around 4 mm. It is important to realise that this is not a simple crack bridged by fibres. However, the behaviour can be viewed in terms of the motion of a simple Griffith crack tied together by the fibres. Fig 7b shows that for very small cracks the toughness will be dominated by the matrix value of K_{1C1}. But when the crack extends, the fibres bridge the crack and K_{1C} increases approximately with $\sqrt{c}$ as indicated theoretically by Rose [15] and McCartney [16] until the fibres begin to break and a limiting value of K_{1C2} is attained. In effect, the material follows two Griffith curves (Fig 8) and jumps from the matrix curve to the limiting curve over a certain range of crack lengths, in this case from crack lengths of 0.1 mm to crack lengths 4 mm.

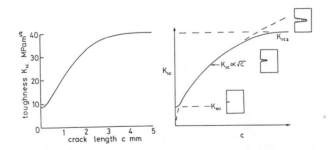

Fig.7. a) R curve behaviour of silicon carbide fibre reinforced
          silicon carbide matrix [2]
       b) Schematic description of composite toughness resulting
          from matrix crack, bridged crack, and broken fibre regime.

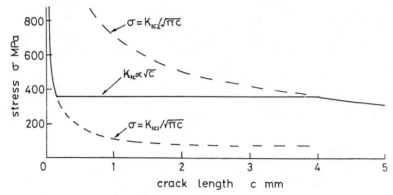

Fig. 8   Composite behaviour viewed as two Griffith curves
bridged by R-curve

Clearly it is only in this crack length range where K_{1C} is increasing
that Weibull modulus can be improved in the toughened material. When K_{1C}
is effectively constant the Weibull modulus cannot rise. Consider, for
example, a matrix with a distribution of maximum flaw sizes from 50 μm to
100 μm. In a test of 10 samples, the Weibull modulus will be 4.7. This
will not change in the material of Fig 7c because the composite shows
Griffith behaviour in this regime. However, if the flaw distribution lies
between 100 and 200 μm, the Weibull modulus will rise to 8 because the
larger flaws will all cause failure at a stress of 350 MPa. If the flaw
distribution lies between 200 μm and 4 mm, the Weibull modulus will be very
large because of the essentially constant failure stress in all the
samples.

CONCLUSIONS

Toughening a ceramic does not necessarily lead to an increase in
Weibull modulus. In fact, toughness itself has no influence on Weibull
modulus for a brittle material. It is R curve behaviour, that is the
increase in toughness for longer cracks, which causes the improvements of
Weibull modulus observed experimentally in zirconia, Dugdale materials and
fibre reinforced ceramics.

REFERENCES

1.   P.F. Becher and G.C. Wei, J. Am. Ceram. Soc. 67, C-267-69 (1984).
2.   P.J. Lamicq, G.A. Bernhart, M.M. Dauchier and J.G. Mace, Am. Ceram.
     Soc. Bull 65, 336 (1986).
3.   M.V. Swain and R.J.H. Hannink Advances in Ceramics 12 (eds. Claussen,
     Ruhle and Heuer) (Am. Ceram. Soc. 1984) p.225.
4.   D.L. Hartsock and A.F. McLean Am.Ceram. Soc. Bull. 63, 266 (1984).
5.   K. Kendall, N.McN. Alford, S.R. Tan and J.D. Birchall, J. Materials
     Res. 1, 120 (1986).
6.   A.A. Griffith, Phil. Trans. R. Soc. Lond. A221, 163 (1920).
7.   D.S. Dugdale, J. Mech. Phys. Solids, 8, 100 (1960).
8.   R.A. Sack, Proc. Phys. Soc. 58, 729 (1946).
9.   J.E. Ritter and R.W. Davidge, J. Am. Ceram. Soc. 67, 432 (1984).
10. W. Weibull, J. Appl. Mech. 18, 293 (1951).
11. K. Kendall, N.McN. Alford and J.D. Birchall, Inst. Ceram. Proc. in
     press.
12. F.J. Burdekin and D.E.W. Stone, J. Strain Analysis 1, 145 (1966).

13. R. Hayami, K. Ueno, I. Kondo and Y. Toibana, Conference on Tailoring Multiphase and Composite Ceramics, Penn. State University July 1985.
14. D.C. Larsen and J.W. Adams <u>Proceedings of the 22nd Department of Energy ATD Contractors Co-ordination Meeting</u>, (Society of Automotive Engineers, Warrendale 1985) p.399.
15. L.R.F. Rose, Int. J. Fracture <u>18</u>, 135 (1982); J. Mech. Phys. Solids, in press (1986).
16. L.N. McCartney, Proc. R. Soc. Lond. 1986 in press.
17. D.B. Marshall, J. Am. Ceram. Soc. <u>69</u>, 173 (1986)

EFFECT OF CHANGES IN GRAIN BOUNDARY TOUGHNESS ON THE STRENGTH OF ALUMINA

ROBERT F. COOK
IBM Thomas J. Watson Research Center, Yorktown Heights, NY 10598

ABSTRACT

Indentation/strength data on a series of Ca-doped aluminas are presented, revealing considerable microstructural influence and increasing fracture resistance during crack propagation. The source of the increasing resistance is considered in terms of a restraining zone of grain-localized ligamentary bridges behind the crack tip. A simple fracture mechanics model is developed to characterize the effect of the restraining ligaments, which also allows the steady state toughness, applicable to large cracks, and the intrinsic strength, applicable to small cracks, to be predicted. The segregation of Ca to the grain boundaries is shown to correlate with decreases in grain boundary toughness and increases in toughness in the large crack limit. These latter two combined effects lead to decreases in intrinsic strength with increasing steady state toughness.

INTRODUCTION

In a recent paper Cook et al used an indentation-strength technique to show that many nontransforming ceramic materials exhibit the property of increasing fracture resistance on crack extension [1]. The strengths of specimens containing dominant indentation flaws were measured as a function of indentation load and significant departures from the usual -1/3 power law dependence observed in many materials, consistent with an increasing crack resistance on extension. A conclusion drawn from this work was that grain boundary properties, as well as grain size, were key to determining the microstructural influences on crack propagation. In the range of polycrystalline aluminas examined the grain boundaries were usually less tough than the matrix grains (single crystal sapphire). However, the influence of the microstructure acted to make the material toughness greater than that of sapphire, to various degrees, as the cracks extended. Cook et al modelled the microstructural influence (somewhat phenomenologically) as a positive decreasing function of crack length to describe this effect.

Knehans and Steinbrech explicitly measured the increases in crack resistance in alumina using the notched beam geometry [2]. By sawing material away from the walls of extended cracks they reduced the crack resistance and showed that the mechanism responsible for toughening must be operating behind the crack tip. They also showed that although the toughening mechanism was distributed over the entire crack it was most effective closer to the tip. Swanson et al have propagated cracks in an alumina exhibiting strong increases in toughness with crack length and observe many interfacial bridges of material spanning the crack walls [3] . The ligamentary bridges are formed by either frictional interlocking of grains or regions of unruptured material and support some of the applied loading, effectively restraining the crack.

The bridging mechanism observed by Swanson et al would appear to well describe the results of Cook et al and Knehans and Steinbrech: as a crack propagates more bridges form, increasing the crack resistance; if material is removed from the crack walls behind the tip bridges are removed, and the crack resistance reverts to a lower value.

Mat. Res. Soc. Symp. Proc. Vol. 78. ⁀1987 Materials Research Society

Here we will use the bridging zone ideas to model the microstructural influence in a simple way which puts the analysis of Cook et al on a more physical basis. The resulting analysis will be used to examine the strength properties of a series of alumina materials as a function of grain size and grain boundary chemistry, providing insight into the relation between toughness and strength.

THEORY

In the indentation-strength method specimens containing indentation flaws made with contact load P are ruptured by an applied stress σ_a under equilibrium conditions. The condition for equilibrium crack propagation is that the net stress intensity factor K (including any contributions from the microstructure, K_μ) equal the intrinsic interfacial (grain boundary) toughness, T_0 :

$$K = K_a + K_r + K_\mu = T_0 \tag{1}$$

where K_a and K_r are the stress intensity factors arising from the applied stress and the residual contact field of the indentation:

$$K_a = \psi \sigma_a c^{1/2} \tag{2}$$

$$K_r = \chi P / c^{3/2} \tag{3}$$

and ψ and χ are crack geometry and elastic/plastic contact parameters, respectively and c is the crack length [4,5] . The toughness of the material is given by

$$T(c) = T_0 - K_\mu(c) \tag{4}$$

where we emphasize that the interfacial toughness T_0 is crack length invariant. To arrive at an appropriate form for $K_\mu(c)$ we must consider the mechanism of microstructural interaction.

Figure 1a is a schematic of the crack paths observed by Swanson et al and in the alumina materials studied here. Close to the crack tip grains bridge the crack, further back from the tip, as the crack profile widens, these grains become separated from the interface, until at some distance behind the tip the grains either rupture or separate completely from matrix. The restraining stress distribution acting over the crack, averaged over many bridging sites, is shown in Fig. 1b. The stress is high close to the crack tip as the full strength of the unruptured grains and frictional interlocking is felt, and diminishes behind the tip. We may represent this stress distribution by a line force σ^* centered a characteristic distance δ behind the tip as shown in Fig. 1c. The stress intensity factor for such a configuration is [6]

$$K_\mu = - 2\sigma^*(c - \delta)/(\pi c)^{1/2}(2c\delta - \delta^2)^{1/2}$$

For use in the strength analysis to follow a more convenient expression for K_μ which has the same functional dependence as that above is

$$K_\mu = - \mu\sigma^*\delta^{-1/2} [1 - (\delta/c)^{3/2}] \tag{5}$$

where μ is numerical constant (approximately $\sqrt{2/\pi}$). Combining Eqs. 4 and 5 we see that in the limit of small cracks ($c/\delta \to 1$) the toughness tends to T_0 and that in the large crack limit ($c/\delta \to \infty$) to another invariant quantity

$$T_\infty = T_0 + \mu \sigma^* \delta^{-1/2} \tag{6}$$

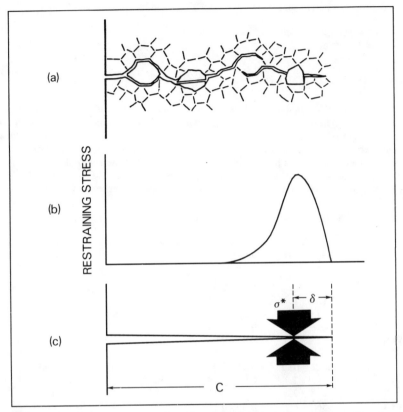

Figure 1. (a) Schematic diagram of crack propagation through a polycrystalline material with grains bridging the crack close to the tip. (b) The restraining stress exerted by the bridges. (c) The model chosen to represent the effect of the restraining stress.

Equation 5 has the same form as that used by Cook et al [1] to describe the microstructural influence, and we may simply rewrite the strength-indentation load expressions derived by them. It is useful to define the quantity

$$P^* = \mu \, \sigma^* \, \delta/\chi \tag{7}$$

characterizing the effect of the microstructural restraining force. Hence in the large flaw limit ($P \gg P^*$) we have the usual $P^{-1/3}$ dependence

$$\sigma_m^P = 3T_\infty^{4/3}/4^{4/3}\psi(\chi P)^{1/3}, \tag{8a}$$

in the small flaw limit ($P \ll P^*$) the tendency to a load independent plateau

$$\sigma_m^0 = 3T_\infty^{4/3}/4^{4/3}\psi(\chi P^*)^{1/3} \tag{8b}$$

and generally

$$\sigma_m(P) = \sigma_m^0 (1 + P/P^*)^{-1/3}. \tag{9}$$

We define an apparent toughness $T^*(P)$ at a given load by analogy with Eq. 8a

$$T^* = T_\infty(\sigma_m P^{1/3}/\sigma_m^P P^{1/3})^{3/4} = T_\infty (1 + P^*/P)^{-1/4} \tag{10}$$

EXPERIMENTAL RESULTS

The materials studied were a series of aluminas, containing three initial levels of calcium doping, heat treated for three different periods, to yield nine materials in all. The average grain size is given in Table I, as the mean intercept length λ measured on the as-fired surfaces. To characterize the segregation of Ca to the grain boundaries Auger electron spectroscopy was performed on fracture surfaces of the materials [7]. The grain boundary concentration of Ca is given in Table I, as the ratio of CaO units to Al_2O_3 units. The test-specimens were in the form of disks 25 mm in diameter and 1.1 mm thick.

Table I - Microstructural and Fracture Parameters of Aluminas

Material	Grain Size λ (μm)	Grain Boundary Concentration (CaO/Al_2O_3)	σ_m^0 (MPa)	P^* (N)	$\sigma_m^P P^{1/3}$ (MPa $N^{1/3}$)
V1	13.8±1.4	0.25±0.04	289	20	784
V2	17.0±1.7	0.21±0.04	221	82	960
V3	48.8±14.4	0.36±0.05	180	920	1751
VC1	12.2±1.3	0.32±0.03	268	39	909
VC2	23.4±3.3	0.42±0.03	201	276	1309
VC3	59.1±10.7	0.34±0.04	182	1203	1936
VCC1	16.3±2.7	0.31±0.03	250	40	855
VCC2	29.0±2.6	0.52±0.11	198	819	1853
VCC3	56.8±6.5	0.69±0.15	172	8709	3539

The testing procedure consisted of indenting the specimens using a Vickers diamond pyramid at contact loads between 2 and 100 N. The specimens were then loaded to failure in a flat on three ball biaxial stressing rig [8]. To ensure inert conditions failure was produced in 50 ms or less and failure loads monitored with a piezoelectric load cell. Failure stresses were calculated by simple plate theory. Some control specimens were not indented prior to the strength test.

Figure 2a plots the mean strength of six - eight tests at the designated indentation loads for the nine materials. The data have been plotted by normalizing the strengths to the plateau strength σ_m^0 and the indentation loads to P^*, obtained by best-fitting the raw data to strength-load responses described by Eq. 9. The resulting fitting parameters are given in Table I. Also given in Table I is the parameter $\sigma_m^P P^{1/3} = \sigma_m^0 P^* {}^{1/3}$ which is proportional to the steady state toughness T_∞ (Eq. 8a). Figure 2b plots the apparent toughness T^* normalized by T_∞, using Eq. 10 and the strength data, against the same normalized indentation load.

Figure 3 shows the strengths of specimens tested either without indentation flaws or which did not fail at indentation flaws. Each symbol represents the mean of approximately five tests. The data are normalized as in Fig. 2, where no indentation was used P/P^* was set to the ratio at which the probability of failure from indentation flaws was expected to be zero.

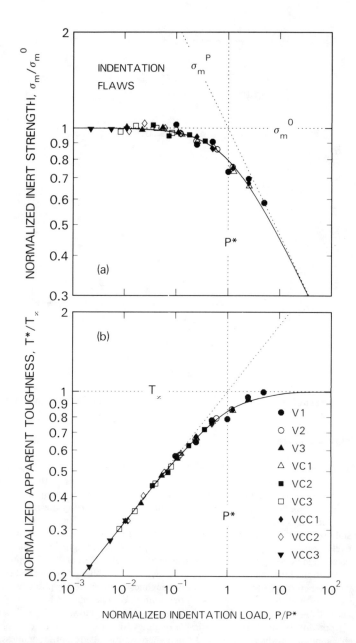

<u>Figure 2.</u> (a) Inert strength and (b) Apparent toughness as a function of indentation load for the aluminas. Dotted lines indicate asymptotic limits (Eq. 8) and solid lines general solutions (Eqs. 9 and 10).

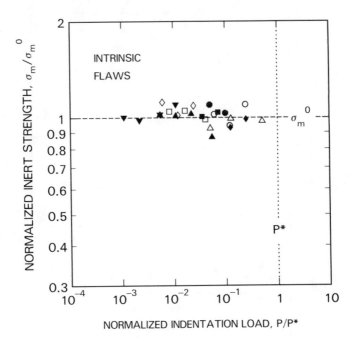

Figure 3. Inert strength of specimens which did not fail from indentation flaws at the loads indicated.

DISCUSSION AND CONCLUSIONS

The agreement between the data and the predicted responses in Fig. 2 suggests that the simple model used here is capable of describing observed strength and apparent toughness behavior as functions of flaw size. In particular we note that specimens tested with indentation loads small in comparison with that characterizing the microstructural restraining force P^* exhibit large changes in apparent toughness T^* with P (Fig. 2b). The strong stabilization of the propagating cracks reflected in the changes in toughness leads to relatively constant measured strengths (Fig. 2a). This strength invariance may be identified as an increasing Weibull modulus, as noted by Kendall et al [9]. To optimize this quality through the bridging mechanism, Eq. 7 suggests that either σ^* be maximized, which presumably depends on the grain boundary toughness and grain morphology and/or δ maximized, which presumably depends on the grain size. The relation between P^* and λ exhibited in Table I probably reflects this latter dependence. However, there is some indication from Eqs. 6 and 8 that all desirable qualities for structural ceramics may not be achievable simultaneously. Maximizing P^* may increase the steady state toughness T_∞ and the degree of flaw tolerance but decrease the plateau strength in the small flaw limit σ_m^0, suggesting that compromise may be necessary. The inverse relationship between the $\sigma_m^P P^{1/3}$ and σ_m^0 values in Table I is an example of this.

Indentation-strength tests may also be capable of predicting the strengths of materials failing from flaws intrinsic to the microstructure. The tendency for the intrinsic strengths to lie on the $\sigma_m = \sigma_m^0$ line in Fig. 3 supports this contention and also indicates that a stabilizing bridging zone is formed during the propagation of cracks intrinsic to the material. The quality of flaw tolerance is highlighted in Fig. 3 by the fact that failure from intrinsic flaws still occurs at indentation loads up to approximately P^*. The probability of failure from an indentation flaw is plotted in Fig. 4 as a function of the normalized load P/P^*. The gradual transition from intrinsic flaw dominated failure to extrinsic (indentation) dominated failure as $P \rightarrow P^*$ is a consequence of the strong stabilization exerted on both types of flaws. In materials with negligible toughness variation (eg. glass) the transition is more abrupt at $P \simeq P^*$.

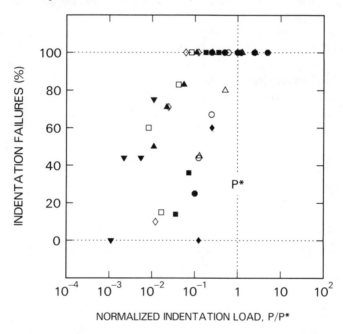

Figure 4. Probability of failure from an indentation flaw as a function of indentation load.

Increases in the restraining force exerted by the bridging zone in the aluminas examined here appear to be associated with the segregation of Ca to the grain boundaries, leading to a similar dependence of the large crack toughness on the Ca segregation (Table I). As it is precisely the high T_∞ materials which exhibit the strongest decreases in apparent toughness with decreasing flaw size (Fig. 2b), we may infer that the segregation of Ca decreased the toughness of the grain boundaries in these aluminas. However, evaluation of T_0 from Eqs. 6 and 7 using the values of T_∞ and P^* requires details of the bridging process (ie specification of either σ^* or δ) which we have excluded from our model. Explicit models of the ligamenting process will allow the determination of the intrinsic interfacial toughness [10], and this will be the subject of a future publication [11]. Nevertheless, the implication is that the segregation of Ca to the grain boundaries, although lowering the grain boundary toughness, increases the tendency or effectiveness of the ligamenting process, thus increasing the steady-state toughness.

206

REFERENCES

1. R.F. Cook, B.R. Lawn and C.J. Fairbanks, J.Am.Ceram.Soc. 68 , 604 (1985)
2. R. Knehans and R. Steinbrech, J.Mater.Sci.Letters 1 , 327 (1982)
3. P.L. Swanson, C.J. Fairbanks, B.R. Lawn, Y-W Mai and B.R. Hockey, J.Am.Ceram.Soc., in press
4. Y-W Mai and B.R. Lawn, Ann.Rev.Mater.Sci. 16 , 415 (1986)
5. B.R. Lawn, in Fracture Mechanics of Ceramics, Vol. 5, edited by R.C. Bradt, A.G. Evans, D.P.H. Hasselman and F.F. Lange (Plenum Publishing Corporation, New York, 1983) p. 1.
6. H. Tada, P.C. Paris and G.R.Irwin, The Stress Analysis of Cracks Handbook (Del Research Corporation, St. Louis, 1973)
7. R.F. Cook and A.G. Schrott, to be published
8. D.B. Marshall, Am.Ceram.Soc.Bull. 59 , 551 (1980)
9. K. Kendall, N.McN. Alford, S.R. Tan and J.D. Birchall, J.Mater.Res. 1 , 120 (1986)
10. R.F. Cook, C.J. Fairbanks, B.R. Lawn and Y-W Mai, J.Am.Ceram.Soc. submitted
11. R.F. Cook, to be published

EROSION OF A SILICON CARBIDE WHISKER REINFORCED SILICON NITRIDE

C.T. MORRISON[*], J.L. ROUTBORT[**] AND R.O. SCATTERGOOD[*]
[*]Dept of Materials Engr., North Carolina State Univ, Raleigh, NC 27695
[**]Matl. Science Div, Argonne National Laboratory, Argonne, IL 60439

ABSTRACT

The steady-state solid particle erosion behavior of hot-pressed Si_3N_4 reinforced with 0, 10, and 20 weight percent SiC whiskers has been investigated at room temperature using angular alumina particles (63 to 270 μm diameter) as the erodent. The impact angle was varied from 30 to 90° and particle velocities were varied from 80 to 140 m/s. These materials were found to be very erosion resistant. However, there was little effect of the SiC whisker reinforcement on either the absolute erosion rate or on the velocity or particle-size exponents for the rates.The lack of a fiber reinforcement effect occurs even though the fiber additions provide an increase in the fracture toughness at high fiber contents. Results on differently processed Si_3N_4 base materials showed that microstructural variations due to processing history have a very significant influence on the erosion resistance of the base (matrix) material. The results substantiate the idea that microstructure plays an important, but not fully understood role in the erosion processes for brittle materials.

INTRODUCTION

Reinforcement of ceramics by high-strength whiskers can lead to an improvement of fracture toughness. In general, fiber reinforcement improves fracture toughness by mechanisms related to fiber sliding, crack deflection or microcrack formation [1]. Indeed, the toughness of Si_3N_4 reinforced with up to 40 volume percent SiC whiskers produced by a Vapor-Liquid-Solid process increased by 50% [2]. This implies that the erosion resistance of ceramics, which depends on the fracture toughness, can be increased by the addition of whiskers. Depending on erosion conditions, increases in erosion resistance of a factor of three or more were observed in Al_2O_3 matrices that had been reinforced with up to 25 weight percent SiC whiskers whose approximate diameter and length were 1 and 30 μm respectively [3]. The improvement in the absolute erosion rates and a decrease in the velocity and size exponents for the steady-state erosion rates were explained on the basis of a modification to the existing erosion models. In this analysis, it was necessary to account for the effect of crack-closing pressure due to fiber pull out on lateral crack propagation [3]. In the Al_2O_3 work [3], the value of velocity exponent ranged from approximately 3 for the base material to 1 for the Al_2O_3 with 25% SiC, in reasonable agreement with the predictions of 2.44 to 1.56 calculated for the transition from matrix to fiber control under quasi-static contact conditions.

The objective of the work reported here was to examine the erosion resistance of another fiber-reinforced ceramic system. The results for SiC reinforced Si_3N_4 reported by Shalek, et. al. [2] demonstrated that fracture toughness increases can be obtained in this system. However, because of the good SiC - Si_3N_4 fiber-matrix bonding, the improvements in toughness appear to be related to crack deflection or microcracking mechanisms rather than the fiber sliding observed with Al_2O_3 matrices. Furthermore, since processing methods can have a significant effect on the performance of ceramics, two different unreinforced Si_3N_4 base materials were tested to assess the possible influence of processing on the matrix properties themselves. Clearly, if processing varies as a result of additives or fiber blending techniques, then processing induced microstructural variations could obscure or even negate the beneficial effects of the fiber reinforcement.

EXPERIMENTAL

Material

The Si_3N_4 - SiC whisker reinforced material was kindly supplied by Dr. Jim Rhodes,

Arco Chemical Corp., Greer, SC. It was produced from a commerical powder to which 3 wt. % MgO and the SiC whiskers were added and blended. The whiskers were produced from rice hulls and were approximately 1 μm in diameter by 30 μm long. They were added to the powder in compositions of 0, 10, and 20 weight percent. A 29 mm diameter disc, 6 mm thick, was produced by hot-pressing at between 1720 and 1750°C at a pressure of 28 MPa in a graphite die with a BN liner in a N_2 atmosphere. For comparison, a Si_3N_4 with 6 wt. % Y_2O_3 was prepared by Dr. J.P. Singh at Argonne National Laboratory (ANL) by hot-pressing in N_2 at the same temperature at a pressure of 22 MPa [4]. X-ray analysis was used to determine that the final compact had the same composition as the initial powders. In both cases the densities were 100% of the theoretical density. A notable difference between the two materials was that there were laminations or striations perpendicular to the hot-pressing direction in all of the material prepared by Arco (with and without fiber aditions). The orgin of the laminations is not clear. The ANL material was free of these laminations and was produced only in the unreinforced condition.

Experimental Technique

The erosion tests were carried out using a slinger-type apparatus described by Kosel et al. [5]. A rotating barrel operated in vacuum is used to accelerate the erodent particles to a velocity fixed by the rotational speed of the barrel. A series of samples having different compositions and different impact angles can be tested during one run. Steady-state erosion rates are determined from plots of the sample mass loss vs. the mass of particles impacted on the sample surface. When the plot becomes linear, the slope determines the steady-state erosion rate. The steady-state erosion rates ΔW are given as g/g unless otherwise noted.

The erodent particles used for the tests were commercial angular-shaped alumina abrasives (Norton Alundum 38) with mean diameters ϕ of 63 and 270 μm. Details on the size distribution are available elsewhere [6]. Particle velocities v of 80, 100, 120, and 140 m/s were used and the impact angles α varied from 30 to 90°. Scanning electron microscopy (SEM) observations were made to characterize the damage morphology of the eroded surfaces.

RESULTS

The steady-state erosion results for the Arco and ANL base Si_3N_4 materials without whiskers additions are shown in Fig. 1. ΔW is plotted logarithmically as a function of the normal component of the impact velocity, v sin α, for $\phi = 63$ and 270 μm. The ANL material was tested only at $\alpha = 90°$. There are certain features that should be noted for these results. First, the Arco data for each particle size fits onto a high impact angle or a low impact angle line (high angles correspond to $\alpha \geq 60°$ and low angles to $\alpha = 30°$). This trend is almost a universal result for brittle solids and it has been observed in a wide-variety of materials such as Si single crystals [7], mullite [8], and in the Al_2O_3 -SiC whisker reinforced composite [3]. The fact that the ΔW values for $\phi = 63$ μm are somewhat higher than those for $\phi = 270$ μm in Fig. 1 does not mean that the smaller particles remove more material per impact because, with ΔW in g/g, the mass removed per particle impact is proportional to $\phi^3 \Delta W$ (it is convenient to report the results in g/g since this is the experimentally measureable quantity). As will be mentioned below, "ideal" brittle materials show a decrease in ΔW values with decreasing erodent particle size and the trends displayed in Fig. 1 must reflect some type of microstructural interaction. With similar velocities and particle sizes, the resistance to erosion of the Arco material is approximately three times greater than the erosion resistance of the ANL material. The erosion rate of the ANL material is equal to that measured by Hockey, et al. [9] when suitably normalized for particle size and velocity. The Arco Si_3N_4 base material is very erosion resistant in this context.

Fig. 2 is a logarithmic plot of ΔW vs. v sin α for the 10% SiC composite material. It can be seen that with the exception of the data point obtained using 270 μm particles at v = 100 m/s and $\alpha = 90°$, the conventional power-law description of erosion $\Delta W \propto v^n$ is valid, as was the case for all of the data in Fig. 1. In contrast to the base material shown in Fig. 1, the low angle upward shift of the data in Fig. 2 is already apparent at $\alpha = 60°$. It should also be noted that while smaller particles have higher ΔW values at normal incidence, the trend in Fig. 2 is reversed for $\alpha = 30$. Plots similar to Fig. 2 were obtained for the 20% SiC composite and the identical observations concerning the particle size trends were made.

The velocity exponents n obtained from the slopes of the lines in Figs. 1 and 2 and similar plots for the 20 wt.% SiC composite are summarized in Table 1. The values are estimated to be significant to about ± 0.3. There are no systematic trends observed in the exponent values.

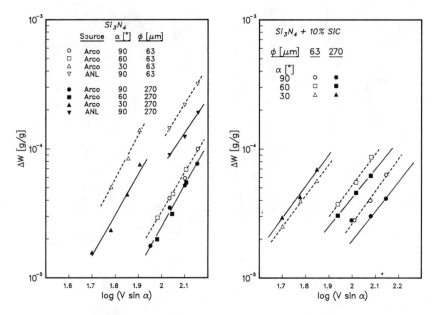

Fig. 1. Logarithm of ΔW vs. v sin α for Arco and ANL Si$_3$N$_4$.

Fig. 2. Logarithm of ΔW vs. v sin α for 10 % SiC composite.

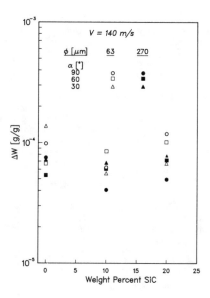

Fig. 3. Logarithm of ΔW vs. % SiC, v = 140 m/s.

TABLE I. Velocity exponents.

	PARTICLE SIZE			
MATERIAL	63 μm diameter		270 μm diameter	
	$\alpha = 30^{\circ}$	90°	30°	90°
Si_3N_4 - ANL	---	2.4	---	2.2
Si_3N_4 - Arco	3.0	2.6	2.6	2.8
Si_3N_4 - 10 wt.% SiC	2.4	2.4	2.5	2.3^{+}
Si_3N_4 - 20 wt.% SiC	2.5	2.2	2.3	2.6^{+}

[+]Exponent taken from v=120 and 140 m/s only (see Fig. 2).

The particle erosion rate $\Delta V \propto \Delta W \phi^3$ (proportional to the mass removed per particle impact) was plotted logarithmically as a function of the impacting particle size ϕ. Only two particle sizes were used in this investigation, but it is well known that that $\Delta V \propto \phi^p$ (see, for example, [7, 8]). Because only two data points were available for each fit, it is difficult to judge the uncertainty in the measured values of p shown in Table II. For an "ideal" brittle material that suffers material loss by the formation of lateral cracks, erosion models predict, and experimental results confirm, that the value p = 3.67 indicating that $\Delta W \propto \phi^{0.67}$ [10, 11]. There is little systematic trend in the particle-size exponents shown in Table II, but they are all significantly less than the theoretical value of 3.67, again suggesting a possible microstructural interaction.

TABLE II. Particle-size exponents.

MATERIAL	$\alpha = 30^{\circ}$	$\alpha = 90^{\circ}$
Si_3N_4 - ANL	---	2.6
Si_3N_4 - Arco	2.6	2.8
Si_3N_4 - 10 wt.% SiC	3.1	2.7
Si_3N_4 - 20 wt.% SiC	3.2	2.4

A plot of ΔW vs. SiC fiber content measured for v = 140 m/s, ϕ= 63 and 270 μm, and α= 30, 60 and 90° for the Arco Si_3N_4 materials is shown in Fig. 3. This figure is typical for all of the erodent particle velocities. In general, the erosion rate is not a sensitive function of the SiC whisker content. There are, however, some slight but reproducible trends. At normal and 30° incidence, ΔW first decreases and then increases as SiC whiskers are added. At 60°, however, ΔW increases monotonically with increasing SiC content. In all cases, the change in erosion resistance with increasing fiber content is rather small, and improvements in erosion resistance of the magnitude obtained in SiC reinforced Al_2O_3 were not observed [3].

Some insight into the erosion processes can be gained from an examination by scanning electron microscopy of the eroded surfaces. Fig. 4 is a SEM micrograph of the Arco Si_3N_4 base material, which was eroded into steady state at normal incidence for 270 μm particles impacting at a velocity of 140 m/s. Fig. 5 is a micrograph of the ANL Si_3N_4 material eroded into steady state under identical conditions. Observations at lower magnifications on the eroded surfaces obtained for a range of conditions showed that, in general, the surface of the Arco material is rather flat and featureless while the ANL material shows considerably more roughness due to intergranular fractures. Neither material exhibits any obvious evidence of the classical lateral crack formation that is so apparent in brittle materials that have a very low fracture toughness [7,8].

Figs. 6 and 7 show SEM micrographs of the 20% and 10% SiC reinforced Arco materials eroded into steady state at normal incidence for 270 and 63 μm particles, respectively. Surfaces eroded at identical conditions in Figs. 4 and 6 for Arco base material and 20 % SiC are similar in overall appearance. Most notable is the fact that SiC fiber pull out is never observed on any eroded surfaces of the fiber reinforced materials. This implies strong fiber-matrix bonding in the SiC - Si_3N_4 system, consistent with the results reported by Shalek, et.al [2].

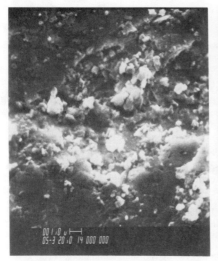

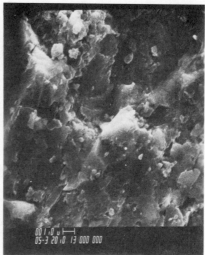

Fig. 4. SEM of eroded surface - Arco Si$_3$N$_4$.
Eroded at α=90°, v=140 m/s and ϕ=270 μm.

Fig. 5. SEM of eroded surface - ANL Si$_3$N$_4$.
Eroded at α=90°, v=140 m/s and ϕ=270 μm.

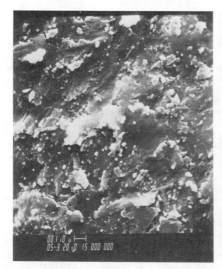

Fig. 6. SEM of eroded surface - 20% SiC.
Eroded at α=90°, v=140 m/s and ϕ=270 μm.

Fig. 7. SEM of eroded surface - 10% SiC.
Eroded at α=90°, v=140 m/s and ϕ=63 μm.

DISCUSSION

Fracture Toughness

The details of fracture toughness and fracture stress measurements on the ANL and Arco materials will be reported elsewhere [12]. Since fracture toughness appears to be a dominant parameter in determining the erosion resistance of brittle materials [13], the results of the fracture toughness measurements pertinent to the present investigation can be summarized as follows. Within the scatter of the measurements, the ANL Si_3N_4 had fracture toughness values that were about 50% higher than the Arco base material. The 20% SiC reinforced Si_3N_4 and base ANL Si_3N_4 materials had similar fracture toughness values. The values of fracture toughness varied between about 4 to 7 MPa-m$^{1/2}$.

Base Materials

Fig. 1 indicates that the Arco base material is about 3 times more erosion resistant than the ANL material despite the fact that the ANL material has somewhat higher fracture toughness. The higher toughness would normally imply more resistance to erosion [13], which is clearly not the case here. However, microstructure is known to play a significant, but poorly understood role in erosion mechanisms [14-16]. It is possible that the grain structures for ANL and Arco Si_3N_4 differed due to processing history. The presence of asymmetric or elongated grains in the Arco Si_3N_4 remains a speculation because attempts at etching grain boundaries were not convincing. This material did, however, show laminations perpendicular to the pressing direction that are indicative of considerable lateral flow during processing (no such laminations occurred in the ANL material). Etching did reveal that the distribution of the glass phase was different in directions perpendicular and parallel to the hot-pressing direction in Arco material. The fracture toughness was also 20% different in directions perpendicular and parallel to the hot-pressing direction. Thus, it is possible that, relative to the ANL material, microstructural differences due to the grain or grain boundary structure significantly improve the erosion resistance in unreinforced Arco Si_3N_4 while at the same time having an adverse affect on macroscopic fracture toughness. Examination of fracture surfaces showed that the grain size in both materials was the order of a few microns, but details related to grain size and grain shape distributions were not determined.

The steady-state erosion surfaces of the ANL material observed by SEM showed considerable roughness due to intergranular fractures while the surfaces for the Arco material were relatively smooth with less intergranular fracture morphology. The improved macroscopic fracture toughness in ANL material could be a result of microcrack formation. The interesting possibility then arises that microcrack shielding or microcrack-induced deflection processes can improve macroscopic fracture toughness, but because of the small scale and near-surface disposition of erosion impact events, the same microcrack phenomena significantly degrade erosion resistance. Stated otherwise, the suppression of microcracking propensity in the Arco material as a result of processing induced microstructural constraint increases the erosion resistance but diminishes the macroscopic fracture toughness relative to ANL material. These microstructural and size scale effects are not well understood and further studies are needed to elucidate the underlying mechanisms.

The shift in ΔW vs. $v \sin \alpha$ for high vs. low impact angles observed for the Arco Si_3N_4 (Fig. 1) has been observed in many nominally brittle materials. In a very brittle material such as glass, this effect has recently been ascribed to the neglect of the tangential component of impact force that will contribute to modes II and III lateral crack propagation [17]. However, the SEM micrographs reveal little evidence of "classical" lateral crack formation in the Si_3N_4 materials, which is consistent with observations on other higher toughness brittle materials [13]. It appears that the very substantial relative upward shift in the low angle data, which is also accompanied by unusually flat erosion rate vs. impact angle curves, may actually reflect a suppression of the normal erosion mechanisms at higher angles. No evidence could be found by SEM observations for the onset of another mechanism that would increase the erosion rates at lower impact angles, for example, the plastic grooving and cutting observed in other ceramic systems [8]. The impact-angle dependent phenomena that produces this suppression is not clear, but it must be related to some type of microstructural constraint that is most effective at normal impact.

The fact that the particle-size exponents in Table II are significantly less than the 3.67 value expected on the basis of lateral crack formation [10, 11] is not surprising in view of the lack of obvious evidence of lateral crack formation in the SEM micrographs. It should also be mentioned that the toughness of the target material is considerably greater than the toughness of the abrasive particles. Therefore, the possibility of erodent particle fragmentation [18] upon impact cannot be overlooked. If fragmentation occurred for the larger particles, the values of p in Table II would have less significance in terms of existing model predictions.

Whisker Reinforced Composites

The results in Fig. 3 indicate that SiC fiber reinforcement has only a minor beneficial effect for improving the erosion resistance of Si_3N_4. This is true even at the 20% SiC fiber content where a noticeable improvement in macroscopic fracture toughness can be measured [12]. The improvement of fracture toughness in SiC reinforced Si_3N_4 appears to be a result of microcracking and crack deflection processes because virtually no fiber pull out is observed by SEM on the fracture surfaces. Thus, the net effect of fiber reinforcement on erosion in SiC reinforced Si_3N_4 composites must be a rather complicated interplay between increases in effective fracture toughness and the possible degradation in erosion resistance due to microcracking per se. The latter effect is rather dramatically illustrated by the results for the Arco and ANL Si_3N_4 base materials. In this same context, processing induced variations in the microstructure can be expected to play an important role in the ultimate performance of fiber-reinforced SiC - Si_3N_4 composites.

The observed low angle shift in the plots of the steady-state erosion rate ΔW vs. v sin α, as illustrated by Fig. 2, can be rationalized in the same manner as the shift observed in the Arco base material, ie., a suppression of the erosion efficiency at high angles due to a microstructural constraint. In fact, the SiC reinforced material already shows a low angle shift appearing at $\alpha = 60^o$, and the erosion rate vs. impact angle curves show unusual behavior with maxima appearing in the range of $\alpha = 50$ to 60°. Increased microcrack formation due to fiber additions could underlie this behavior by interaction with the as yet unknown microstructural constraint. In contrast to the results reported here, SiC reinforced Al_2O_3 composites show a very different behavior wherein the addition of fibers diminishes and eventually eliminates the low angle shift in erosion rates [3].

The results for SiC reinforced Si_3N_4 matrices can be compared further to earlier results for SiC reinforced Al_2O_3 matrices where significant improvements in both macroscopic fracture toughness and erosion resistance can be obtained [3]. In the SiC - Al_2O_3 system, the fracture toughening mechanism appears to be due primarily to fiber sliding. Extensive fiber pull out was observed by SEM on the erosion surfaces, thus indicating that fiber pull out is still a viable toughening mechanism for the local impact conditions imposed by erosion impacts. The toughening in this case is apparently not affected by size scale effects. The erodent particle size and particle velocity exponents for the steady-state erosion rate also showed systematic decreases with increasing fiber content, consistent with theoretical predictions based on closing pressure effects due to fiber sliding. No systematic trends were observed for the particle or velocity exponents for the steady-state erosion rates measured in this investigation. This is in agreement with the lack of fiber sliding observed on the erosion surfaces.

In summary, the erosion resistance of Si_3N_4 matrices is not significantly affected by addition of SiC reinforcing fibers even though an improvement in macroscopic fracture toughness can be obtained. The strong bonding between SiC fibers and Si_3N_4 matrices prevents fiber sliding and therefore microcrack and crack deflection processes must be dominant for fracture toughening mechanisms in this system. Because of poorly understood microstructural interactions and associated size scale effects, these toughening mechanisms do not improve the resistance to the localized, near-surface fracture processes operative during solid-particle erosion impact events. Hence, erosion resistance is not significantly improved. Observations on differently processed, unreinforced Si_3N_4 base materials show that the relative erosion resistance can be severely degraded by microcracking processes that can improve the overall macroscopic fracture toughness.

214

ACKNOWLEDGEMENTS

The authors are extremely grateful to Dr. J.F. Rhodes of Arco Chemical Co, Greer, SC who supplied all of the fiber reinforced material, which was made especially for this investigation. We are also pleased to acknowledge Dr. J.P. Singh of Argonne National Laboratory who provided us with another high-quality base Si_3N_4 used as a comparison material. The authors have also benefited from fruitful scientific discussions with both of these gentlemen. This work was supported by the U.S. Department of Energy, Basic Energy Sciences, Material Science, under Contract W-31-109-Eng-38 and DE-FG05-84ER45115. The work that was performed at Argonne National Laboratory by one of the authors (C.T.M.) was under a thesis program administered by the Argonne Division of Educational Programs.

REFERENCES

[1]   R. W. Rice, Cer. Eng. Sci. Proc. 2, 661 (1981).
[2]   P.D. Shalek, J.J. Petrovic, G.F. Hurley and F.D. Gac, Am. Ceram. Soc. Bull. 65, 351 (1986).
[3]   M.T. Sykes, R.O. Scattergood and J.L. Routbort, Composites, in press.
[4]   J.P. Singh, to be published
[5]   T.H.Kosel, R.O. Scattergood and A.P.L. Turner, in Wear of Materials, edited by K. C. Ludema, W.A. Glaeser and S.K. Rhee (ASME, New York, 1979) p. 192.
[6]   J.L. Routbort, R.O. Scattergood and A.P.L. Turner, Wear 59, 363 (1980).
[7]   J.L. Routbort, R.O. Scattergood and E.W. Kay, J. Am. Ceram. Soc. 63, 635 (1980).
[8]   C. T. Morrison, J.L. Routbort and R.O. Scattergood, Wear 105, 19 (1985).
[9]   B.J. Hockey, S.M. Wiederhorn and H. Johnson, in Fracture Mechanics of Ceramics, edited by R.C. Bradt, D.P.H. Hasselman and F.F. Lange (Plenum Press, New York, 1977) p.379.
[10]   A.G. Evans, M.E. Gulden and M. Rosenblatt, Proc. R. Soc. London Ser. A 361, 343 (1978).
[11]   S.M. Wiederhorn and B.R. Lawn, J. Am. Ceram. Soc. 62, 66 (1979).
[12]   J.P. Singh, K.C. Goretta, J.L. Routbort and J.F. Rhodes, to be published
[13]   S.M. Wiederhorn and B.J. Hockey, J. Mat. Sci. 18, 766 (1983).
[14]   J.L. Routbort and R.O. Scattergood, J. Am. Ceram. Soc. 63, 593 (1980).
[15]   J.L. Routbort and Hj. Matzke, J. of Matl. Science 18, 1491 (1983).
[16]   J.E. Ritter, Matl. Science and Engr. 71, 195 (1985).
[17]   S. Srinivasan and R.O. Scattergood, J. of Matl. Science, in press
[18]   G.P. Tilly and W. Sage, Wear 11, 123 (1968).

COMPARISON OF MATRIX VARIATION IN NICALON
(SiC) FIBER REINFORCED COMPOSITES

W. L. Johnson, III, and R. G. Brasfield

Morton Thiokol, Inc.
P.O. Box 241
Elkton, Maryland 21921

ABSTRACT

The high-temperature behavior of NICALON fabric is discussed. Three fabric reinforced composites are described that use NICALON fabric to take advantage of its oxidation and erosion resistance. The matrix materials used were phenolic resin, pyrolytic carbon and beta-silicon carbide. Each composite was subjected to an erosive environment at 2755 K. Scanning electron microscopy (SEM), transmission electron microscopy (TEM), and laser Raman microprobe analysis (LRMA) were used to characterize the raw materials, the unfired composites, and the erosion tested composites. These analyses concentrated on the fiber/matrix interaction and the effect of erosion on each component of the composite.

INTRODUCTION

Erosion at service temperature is a factor that can limit the usefulness of a refractory composite. Silicon carbide has received considerable attention in the aerospace industry because it is a refractory material that shows resistance to oxidation and erosion.

A process analogous to that commonly used to synthesize carbon and graphite fibers, pyrolysis of an organic polymeric substrate, has been developed by Yajima [1] to make the silicon carbide fiber NICALON. NICALON is produced by the Nippon Carbon Company as an eight-harness satin cloth suitable for making two-dimensional (2D) laminated composites. In order to evaluate this material, composites having three different matrices were fabricated: silicon carbon-phenolic resin (SiC-P), silicon carbide-pyrolytic carbon (SiC-C), and silicon carbide-silicon carbide (SiC-SiC).

EXPERIMENTAL PROCEDURE

NICALON 8-harness satin weave cloth, phenolic resin, thermal black-filled phenolic resin, and SiC powdered abrasive were used to prepare the fiber-resin plastic stage composites for this study.

The SiC-P and SiC-C composites were made from NICALON cloth impregnated with phenolic resin containing 14% SiC powder as filler. SiC-C composite was made from the SiC-P composite by carbonization/pyrolysis of the material to 1273 K followed by pyrolytic carbon vapor infiltration. SiC-SiC composite was made using a process similar to that used for the SiC-C material, except that the carbonized/pyrolyzed panel material was densified with SiC using the CVD process. These composites are described further in Table I.

TABLE I
EROSION RESISTANT REFRACTORY COMPOSITES

Composite Type	Fiber	Matrix	Max HTT* (K)
SiC-P	NICALON	Phenolic + SiC powder	423
SiC-C	NICALON	Glassy carbon +SiC powder +Pyrolytic carbon	1473
SiC-SiC	NICALON	Glassy carbon +Thermal black +Vapor deposited SiC	1773
*HTT = heat treatment temperature			

SEM and LRMA were performed at McCrone Associates, Chicago, Illinois. Each of the composite specimens was characterized by SEM using secondary as well as backscattered electron imaging.

The I.S.A. Molecular Optics Laser Examiner interfaced to an IBM PC computer was used to measure the Raman spectra for this study. LRMA was performed using the 514.5 nm excitation line of an argon ion laser. Due to weak signals, data were usually acquired using approximately 30mW laser power and a spot size of ~5 microns.

Samples of each of the above composites were subjected to erosion testing in a 2755 K gas stream (Fig. 1) having an initial Reynolds' number of 2×10^6 and a mass flow rate of 182 g/sec. Flat washers of each composite were machined from flat laminate panels as shown in Fig. 2.

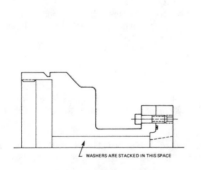

Fig. 1. Hardware for Erosion Testing of Composite Washers

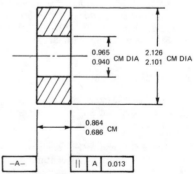

Fig. 2. Composite Washer Design for the Erosion Testing of Refractory Composites

RESULTS AND DISCUSSION

Refractory Silicon Carbide Fabric

NICALON fiber is made by the thermal degradation of polycarbosiloxane fiber that is made by a synthesis involving three polymeric stages: polydimethyl-silane, polycarbosilane, and polycarbosiloxane. Initially, dimethyldichlorosilane is reacted with sodium metal to yield polydimethylsilane. This polymer is heated 14 hours at 723 to 743 K under 100 kPa pressure to yield polycarbosilane. At this stage fibers are hot-spun at 613 K. The fibers are rendered infusible by oxidizing them at 463 K in air to produce polycarbosiloxane fibers. SiC fibers are produced directly from the polycarbosiloxane fibers by pyrolysis at 1473 to 1573 K in an inert atmosphere or vacuum. From the typical chemical analysis of NICALON shown in Table II, it is evident that a substantial portion of the fiber is not SiC. After prolonged heat treatment at 1473 K, the fiber remains substantially undegraded [2].

TABLE II
COMPOSITION OF NICALON FIBERS

	Si (wt. %)	C (wt. %)	O (wt. %)
Typical Composition	54.3	30.0	11.8
After exposure to 1473 K for one hour in air	55.5	30.4	11.6

Upon examination by SEM, the most notable feature of the NICALON fiber surface is that it is smooth, like fiberglass, and nearly circular in cross section (Fig. 3). In addition, no charging effects were observed in the electron micro-scope; therefore, the fiber is deduced to be electrically conductive. The brittle nature of this material is shown by the conchoidal appearance of fiber fracture. Laser Raman microprobe analysis using 458 nm laser excitation was used to characterize the longitudinal fiber surface. The striking feature of the spectrum is that only the spectrum of the nongraphitic carbon was measurable (Fig. 4). Even a change to 514.5 nm laser excitation did not result in a SiC spectrum.

NICALON fiber is black in the "as received" condition. Following heat treat-ment 1283 K, it appears blue in reflected light. Examination of the sample with tungsten light and later by a fluorescent light indicates that the observed color is due to an interference phenomenon caused by a fiber coating of approximately a few thousand angstroms in thickness.

Charging effects were observed when heat-treated NICALON was examined by SEM. This indicates that heat treatment removed the conductive layer from the surface of the fiber. The fibers have occasional surface areas which show spots or pits (Fig. 3). A single laser Raman microprobe scan was unable to detect SiC, but nongraphitic carbon was shown to still be present even after 8 hours at 1283 K in air (Fig. 5). This implies that the carbon is protected by a coating that is transparent to Raman microprobe analysis. Transmission electron microscopy of both the fired and unfired NICALON fibers indicates the fiber consists of

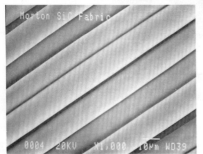

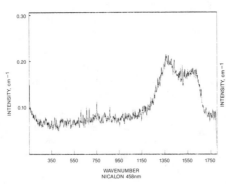

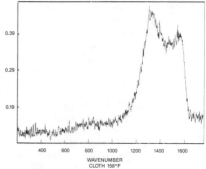

Fig. 3. As-Received and Heat-Treated NICALON

Fig. 4 Raman Spectrum
of NICALON Fiber

Fig. 5. Raman Spectrum
of NICALON Fiber Heated
in Air at 1283 K for 8 Hours

amorphous and crystalline regions of silicon carbide. Diffraction patterns for
these two types of regions were obtained using selected area electron diffraction
(SAED). Table III implies that the silicon carbide possesses the beta structure.

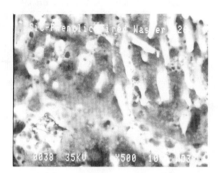

Fig. 6. Unfired and Fired 2D SiC-P Composite

Fig. 7. Unfired and Fired 2D SiC-C Composite

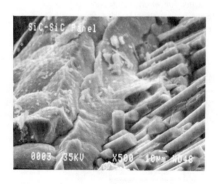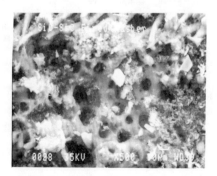

Fig. 8. Unfired and Fired 2D SiC-SiC Composite

TABLE III
X-RAY DIFFRACTION POWDER PATTERN FOR B-SiC COMPARED TO THE
ELECTRON DIFFRACTION PATTERN FOR NICALON FIBER

NICALON (nm)	β-SiC (nm)
0.251	0.252
0.207	0.218
0.159	0.154
0.159	—
0.135	0.131
0.122	0.125
0.111	0.109
0.099	0.100
0.090	0.097
—	0.089
—	0.084
0.079	—

Erosion Resistant Composites

Erosion at service temperature is a factor which can limit the applications for a refractory composite. In a quantitative sense, the erosion rate is a measure of material loss in terms of thickness per unit time. This experimentally determined quantity is dependent upon two overall processes:

. Chemical - Sublimation, melting, reaction with hot gasses, liquids, and solids (corrosion), thermal decomposition

. Mechanical - Thermal shock, thermal coefficient of expansion mismatch, abrasion (erosion)

The results obtained from erosion testing of SiC-P, SiC-C, and SiC-SiC composite are presented in Table IV. In order to evaluate the effect of high-temperature exposure, the composite macrostructure was examined by SEM. Before- and after-high temperature exposure SEM micrographs for each composite are shown in Figs. 6, 7, and 8. A common feature of these composites is that they all have melted material on the fired surface of the composite. This material is most likely SiO_2 (m.p. 1983 K) since carbon and SiC have higher melting points (>2973 K and >3773 K) [3].

TABLE IV
EROSION TEST DATA NICALON COMPOSITES

Composite	Density (g/cc)	Erosion Rate (mm/sec)
SiC-P	1.7	0.014
SiC-C	1.8	0.018
SiC-SiC	2.0	0.0028

Liquid SiO_2 acts to protect the underlying composite to a degree. The oxidation kinetics are determined by the rate of diffusion of O_2 through the SiO_2 layer. During oxidation, CO is formed at the SiC-SiO_2 interface and its desorption from the interface can modulate the rate of oxidation [4, 5]. Of the matrices used to make the three composites, phenolic resin and pyrolytic carbon have the least oxidation resistance and are easily eroded. The ease of matrix erosion results in the exposure of NICALON fiber tips. Conversion of SiC solid to SiO_2 liquid results in a "pencil-pointed" fiber tip. Arrays of fibers exhibiting this phenomenon are shown in Figs. 6 and 7. This feature is not present in the fired SiC-SiC composite (ref. Fig. 8) since the SiC matrix does not erode like phenolic resin and pyrolytic carbon. As a consequence, the NICALON fibers are not exposed to the hot gas flow as they are in the SiC-P and SiC-C composites.

Raman microprobe analysis of these composites yielded primarily evidence of carbon in the NICALON fiber. When the composites were subjected to thermal shock during erosion testing, the type of carbon present in the NICALON fiber changed from a non-graphitic carbon to a diamond-like carbon (Figs. 9 and 10). Evidence of a weak Raman spectrum of β-SiC is seen in the 900 to 1000 cm^{-1} regime.

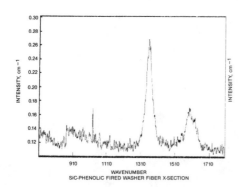

Fig. 9. Raman Spectrum of 2D SiC-P Fired Washer-Fiber Cross Section Surface

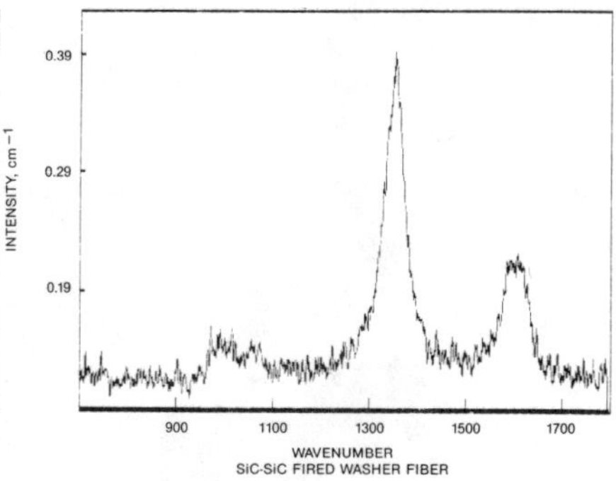

Fig. 10. Raman Spectrum of 2D SiC-SiC Washer-Longitudinal Fiber Surface

CONCLUSIONS

The matrix of an erosion resistant fiber reinforced composite acts to protect fibers. When the matrix is easily lost during erosion tests, the unprotected NICALON fibers develop a pencil-point shape. However, silicon dioxide glass believed to be present on the surface of NICALON fibers can provide protection against further oxidation.

REFERENCES

1. C. H. Andersson and R. Warren, Composites, 15, 17 (1984).

2. T. J. Clark, R. M. Arons, J. B. Stamatoff, and J. Rube, Ceram. Eng. Sci. Proc. 6, 576 (1985).

3. G. Simon and A. R. Bunsell, J. Mater. Sci. 19, 3649 (1984).

4. S. C. Singhal, "Oxidation of Silicon-Based Structural Ceramics," in Properties of High Temperature Alloys (with Emphasis on Environmental Effects, edited by Z.A. Foroulis and F. S. Pettit (Electrochemical Society, Inc., Princeton, NJ, 1977), pp. 697-712.

5. J. W. Hinze, W. C. Tripp, and H. C. Graham, "The High Temperature Oxidation of Hot-Pressed Silicon Carbide," in Mass Transport Phenomena in Ceramics, edited by A. R. Cooper and A. H. Hener, (Plenum Press, 1975).

INTERFACIAL CHARACTERIZATIONS OF FIBER-REINFORCED SiC COMPOSITES
EXHIBITING BRITTLE AND TOUGHENED FRACTURE BEHAVIOR

M.H. RAWLINS, T.A. NOLAN, D.P. STINTON AND R.A. LOWDEN
Oak Ridge National Lab., Oak Ridge, TN 37831

ABSTRACT

A process has been developed for the fabrication of a ceramic composite
of SiC fibers in a chemical vapor infiltrated (CVI) SiC matrix. Early specimens produced by this technique exhibited nonuniform fracture toughness
behavior; regions of brittle fracture with no fiber pullout were interspersed
with regions of good toughness where fiber pullout predominated. Microscopic
and spectroscopic analyses of fiber surfaces revealed that this behavior may
have been related to a thin, discontinuous layer of predominantly silica
which, when present, prevented tight bonding of fibers and CVI matrix.
Consequently, efforts to control interfacial bond strengths and enhance fracture
toughness via fiber pretreatment with CVI overcoatings have had mixed success
depending upon the overcoating specie. Chemical and microstructural characterizations by analytical electron microscopy of these composites are presented
and correlated with composite mechanical property data.

INTRODUCTION

Toughened ceramic composites consisting of fibers or whiskers dispersed
in a ceramic matrix are being developed as a potential solution to some of
today's high temperature materials needs. The fabrication of these composites by a chemical vapor infiltration (CVI) process, developed at the Oak
Ridge National Laboratory, has been reported and is the process used in this
work. [1-3] Specifically, the subject ceramic composites are composed of
Nicalon* (chemically a Si-oxycarbide) fibers in a vapor deposited SiC matrix.
These composite materials must exhibit high temperature strengths comparable
to those of conventional ceramic materials but should also possess improved
fracture toughness. Since the reinforcing fibers are the primary load
carriers in typical composites, of particular importance is the role of the
interface as an energy dissipator during crack propagation. When modest
adhesion exists between fibers and matrix, fiber movements relative to the
matrix are allowed thus inhibiting crack propagation through the composite.
A visual gauge of fracture toughness is the amount of fiber pullout from the
matrix in a failure test indicative of the preferentially weak interface bond.
A significant problem in the material produced by infiltration has been
the lack of reproducibility of the mechanical properties both between specimens and, to a lesser degree, between samples cut from the same specimen,
with strengths varying from 70 to 480 MPa (10 to 70 ksi). This problem was
correlated with inconsistencies in fracture behavior within a specimen;
regions of brittle fracture with no fiber pullout were interspersed with
regions of good toughness where fiber pullout predominated. These observations resulted in chemical and microstructural characterizations of the
fiber surfaces and the fiber-matrix interface and subsequent efforts to
modify the interface region. The results of these efforts is the topic of
this paper.

*Nicalon, Nippon Carbon Co., Tokyo, Japan.

EXPERIMENTAL PROCEDURE

Briefly, fibrous preforms in the shape of disks [45 mm (1.75 in.) diameter x 13 mm (0.5 in.) thick] were fabricated by stacking and compressing multiple layers of a plain-weave Nicalon cloth in a graphite holder (alternate layers oriented at 0-30-60° intervals), yielding a body containing approximately 40 vol.% fibers. The preforms were heated to 1100°C in argon during which time the vinyl acetate sizing was burned off. Deposition of the matrix SiC occurred via the decomposition of methyltrichlorosilane (CH₃SiCl₃) gas which was infiltrated at approximately 1100°C.

Two other composites with modified interfaces were prepared similarly except: in one batch, the fibers in the preforms were precoated with a thin (~1 um) layer of pyrolytic carbon from an isothermal decomposition of infiltrated propylene; and in the other, the fibers were precoated with a thin layer of silicon from an isothermal decomposition of infiltrated silane. It was hoped that the precoats would function to weaken and make uniform the interface bond, and particularly, with the silicon precoating, it was hoped that this would stabilize and protect the Nicalon fibers during matrix infiltration.

Samples for flexure strength testing were cut from the top, middle and bottom of the composite disk. Flexural strengths of various composite specimens were measured at room temperature with four-point loading at a crosshead speed of 0.52 mm/min with an inner span of 6.4 mm and an outer span of 19.1 mm. Fracture toughness was measured using the same four-point conditions using the single-edge notch beam technique.

Scanning electron microscopy (SEM) of fibers and composites was performed using a JEOL JXA 840 and an ISI DS 130 for high resolution work. Transmission electron microscopy (TEM) was performed on dimpled and ion-thinned specimens of composite using a JEOL JEM 2000 FX TEM. Scanning Auger microscopy (SAM) and spectroscopy was performed using a PHI 660 SAM.

RESULTS AND DISCUSSION

The combination of microstructural observations associated with mechanical test results lead early in this program to the conclusion that the Nicalon fiber surfaces were potentially causing nonuniform composite behavior and that the interface would have to be modified to improve fracture toughness. Thus, this discussion is divided into three parts: (1) characterization of composites with no intentional interface modification, (2) characterization of Nicalon fibers prior to CVI, and (3) characterization of composites having fiber surface modifications.

Composites With No Intentional Fiber Surface Modification

A very important factor affecting the strength of the composite material is the bond between the fiber and the matrix. If the bond is too weak, the load does not transfer from the matrix to the fibers and there is no reinforcing. If the bond is too strong, the result is a brittle fracture typical of non-reinforced ceramics. As was discussed in the introduction this composite-type behavior was not obtained consistently for composites with unmodified interfaces. This inconsistency in strength is hypothesized to be related in part to differences in the interfaces, possibly a result of inconsistent surface characteristics of the Nicalon fibers.

Figure 1 is a SEM of a typical fracture surface of the SiC-SiC composite resulting from a flexure test. Considerable variation in fracture toughness is observed from region to region on the fracture surface of a single sample. In the region marked "A" in this figure, the interface bond was such that brittle behavior predominated and a smooth, flat fracture perpendicular to the fiber axes was created. Nearby, there has been considerable pullout of

Figure 1. SEM of a typical fracture surface showing regions of brittle fracture labeled "A" combined with fiber pullout regions labeled "B".

Figure 2. SEM of a fiber surface from a region of pullout. The arrows indicate the thick interfacial film on the fiber surface. The SiC matrix, labeled "M", is detached in the region of the film.

fibers from matrix indicative of improved fracture toughness. Higher magnification SEM of the surfaces of fibers and matrix casts from regions of fiber pullout revealed the presence of a film separating fibers from matrix. This film often appeared to be comprised of at least two layers with a thin, often discontinuous layer next to the fiber and a thicker continuous layer over the first. The matrix SiC was then deposited onto this thick layer. An example of a relatively thick film, is shown in Figure 2; the film remains on a fiber after fracture. The brittle fracture regions appear not to have a film resolvable by SEM. The lack of fiber and cast surface on which to observe films, however, may preclude the observation of a thin film at this interface.

In order to more thoroughly characterize these films and the interface in general, suitable prepared samples were observed by transmission electron microscopy. In Figure 3a, a low magnification view of a typical interface shows the presence of a film with a discontinuous region next to the fiber. In Figure 3b, a higher magnification on another interface, an even more complex structure is visible in which a discontinuous appearing layer is clearly visible as well as a thicker layer. Electron microdiffraction revealed the thicker layer to be very nearly amorphous with rings too diffuse to measure. X-ray energy dispersive analysis revealed that the thicker layer was primarily a silicon compound. The thin layer was also nearly amorphous; its thinness and proximity to the surrounding silicon compounds precluded unambiguous elemental analysis. Figure 3c shows an interfacial region with little or no film present at the interface. Even in this region, however, the interface appears discontinuous.

Nicalon Fiber Characterization

As was described in the experimental section, the fiber cloth is first subjected to a heat treatment in argon to remove impurities and sizing from the fiber surface. It is known that these fibers contain both more carbon than needed for stoichiometric SiC and oxygen; the chemical formula is some-times presented as $Si_3C_{4+3y}O_{1+3x}$ where x and y are deviations from Si_3C_4O composition.

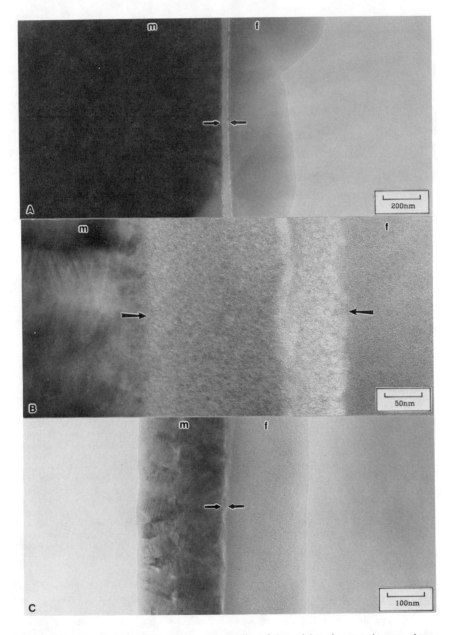

Figure 3. Transmission electron micrographs of ion-thinned composite specimens showing thick, multilayer interfacial films (A and B) in some regions and a very thin interfacial layer in other regions of the same SiC–SiC specimen (C). Fiber and matrix are denoted f and m, respectively, and the interface region is flanked by arrows.

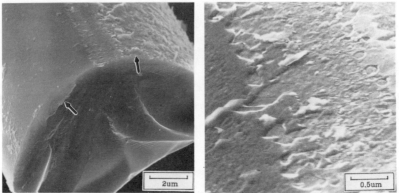

Figure 4. A typical Nicalon fiber after a heat treatment in argon as viewed by SEM at two magnifications. A discontinuous surface film covers much of the fiber.

The heat treatment may also be expected to allow diffusion of both carbon and oxygen from the fiber. The instability of Nicalon fibers at high temperatures has been reported previously.[4-6] It has been shown that various chemical species can diffuse out of the fibers resulting in the evolution of gaseous SiO and CO at temperatures above 1200°C. It was hypothesized that a surface film might remain after the heat treatment either from surface impurities or from reactions involving species diffusing out of the fibers.

In order to determine if a surface film existed prior to the matrix infiltration, samples of furnaced fibers were analyzed using SEM and SAM. In Figure 4 two views of a typical fiber surface reveal that a surface film is present; this film is discontinuous in this view and other regions and other fibers had varying amounts of surface film. Auger electron signal depth profiles for elements of interest were performed on several locations on furnaced filaments. Figure 5 shows the results of a typical profile. Clearly there is a silicon and oxygen rich layer on the surface of the fiber. Auger survey

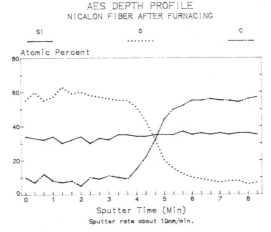

Figure 5. Auger electron depth profiles of the heat treated fiber surfaces reveal a surface film of silica.

scans before and after depth profile sputtering revealed no other elements. (Note that analysis of as received unfurnaced fibers showed considerable surface carbon and no oxide layer.)

In summary, electron microscopy has revealed that in regions of composite exhibiting good fracture toughness the fibers are separated from the matrix by a very fine microcrystalline, porous and discontinuous silicon-containing film. Electron microscopy and Auger spectroscopy show the fibers to be covered in some regions with a film of primarily silicon and oxygen. It is concluded that this silica film, formed by the oxidation of the fiber surface, is responsible for the regions of fracture toughness in this composite.

Composites with Modified Surfaces

In order to produce more uniform behavior and to possibly protect the Nicalon fibers from high temperature degradation, several chemical vapor deposited coatings have been applied to the fibers in the preform prior to SiC infiltration. Two of these coatings have been characterized using electron microscopy. In Figure 6 a fracture is shown of a composite with about one micron of carbon on the fiber surfaces; the carbon was deposited by the isothermal decomposition of propylene. This composite did indeed have more uniform mechanical properties and brittle regions were not observed on specimen fractures. Figure 6 also shows an ion thinned specimen using TEM. High resolution microscopy and electron diffraction revealed that the carbon has a turbostratic structure with the graphitic layer planes oriented to some degree parallel to the fiber surface. There is still evidence of the silica layer between the carbon and the fiber. In another experiment silicon was deposited on the fiber surface by the thermal decomposition of SiH_4 prior to SiC deposition. The composite thus produced was extremely brittle and had practically no fiber pullout. Figure 7 shows a typical fracture surface and also a TEM of the interfacial region. The TEM reveals that the silicon metal is well crystallized. At the time of this report only a few interfaces in this material have been studied using TEM. The ones that have been observed showed no obvious silica layer; this combined with the uniformly brittle behavior suggests that deposition of silicon may have resulted in the removing or chemically changing the oxide film.

CONCLUSIONS

Ceramic fiber-ceramic matrix composites were fabricated by a CVI process and were found to exhibit inconsistent mechanical behavior even within the same specimen. Examination of fracture surfaces by SEM revealed an interfacial film which could be found associated with areas of fiber pullout. This interfacial layer was shown by TEM to be highly variable in thickness, ranging from a few nanometers up to a least 200 nm, and multilayered in some areas with a porous layer next to the fiber surface and a more dense layer covering the porous layer. Auger spectroscopy indicated that this film was silica and was formed during the heat treatment in argon of the fibers prior to infiltration of the matrix-forming gas. This observation is in agreement with the work of others.

The addition of precoatings on the fibrous preforms to enhance composite toughness by making the interface more uniform has shown mixed results. A pyrolytic carbon coating greatly enhanced composite toughness. A silicon precoat, applied to improve toughness and potentially chemically stabilize the fibers at high temperatures, resulted in a composite with little fracture toughness. This was apparently the result of the destruction of the silica layer combined with the inherent brittle nature of the crystalline silicon metal.

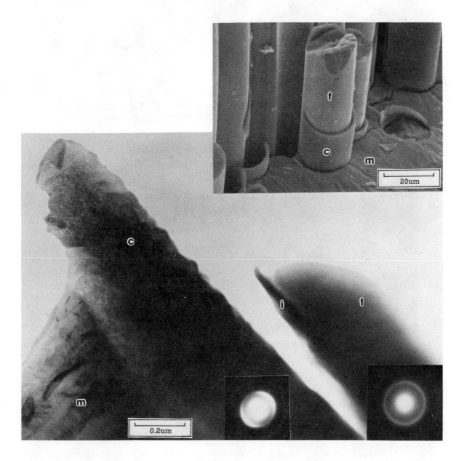

Figure 6. SiC-C-SiC composite by TEM and SEM (upper right). Matrix, carbon, interface layer and fiber are labeled m,c,i and f respectively. Small insets show microdiffraction patterns of the carbon and fibers.

230

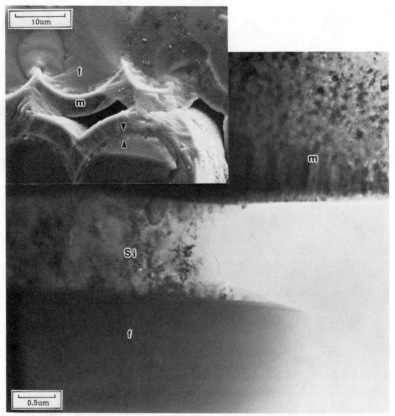

Figure 7. SiC-Si-SiC composite viewed by TEM and SEM (upper left) showing matrix, silicon and fiber (marked m, Si, f respectively). No silica film is observed.

REFERENCES

*Research sponsored by the U.S. Department of Energy, AR&TD Fossil Energy Materials Program, under contract DE-AC05-84OR21400 with Martin Marietta Energy Systems, Inc.
1. A.J. Caputo, W.J. Lackey, "Fabrication of Fiber-Reinforced Ceramic Composites by Chemical Vapor Infiltration," ORNL/TM-9235, Oak Ridge National Lab, Oak Ridge, TN, October 1984.
2. A.J. Caputo, W.J. Lackey, "FAbrication of Fiber-Reinforced Ceramic Composites by Chemical Vapor Infiltration," Ceram. Eng. & Sci. Proc., 5 [7-8] 654-67 (1984).
3. A.J. Caputo, W. J. Lackey, D.P. Stinton, "Development of a New, Faster Process for the Fabrication of Ceramic Fiber-Reinforced Ceramic Composites by Chemical Vapor Infiltration," ibid., 6 [7-8] 694-706 (1985).
4. T. Mah et al., "Thermal Stability of SiC Fibers," J. Mater. Sci. 19, 1191-1201 (1984).
5. L.C. Sawyer et al., Ceram. Eng. & Sci. Proc., 6 [7-8] 567-75 (1985).
6. T.J. Clark et al., Ceram. Eng. & Sci. Proc., 6 [7-8] 576-88 (1985).

ANALYSIS BY SIMS AND BY EELS-EDX IN A STEM

OF SiC FIBERS REINFORCED COMPOSITES.

M. LANCIN*, F; ANXIONNAZ*, M. SCHUHMACHER*, 0. DUGNE*, P. TREBBIA**
*Laboratoire de Physique des Matériaux, CNRS, 1 Pl. A. Briand,92195 Meudon - Cedex, France,
** Laboratoire de Physique des Solides, Université de Paris-Sud, 91405Orsay - Cedex, France.

ABSTRACT:

A series of Nicalon SiC fibers (202) coated by CVD with C and SiC were analysed by Secondary Ion Mass Spectrometry (SIMS). Results (i) give the composition of the material, (ii) show its heterogeneity of composition, and (iii) demonstrate the limit of application of the method for the fibers analyses. SIMS and Electron Energy Loss Spectroscopy (EELS) analyses of one coated fiber segment were done. When elements are detected by both techniques, reasonable agreement is observed between the results. SiC - CVD deposits are not uniform: they are rich in carbon close to the fiber and tend towards stoechiometry after a few microns. The coated fibers are rich in oxygen and in free carbon. Analyses of a glass-ceramic matrix/ Nicalon SiC fiber (202) composite were performed using SIMS, EELS and Energy Dispersive X-Ray Spectroscopy (EDX).

INTRODUCTION:

Materials properties depend on the microstructure and composition of interfaces. Recently, Brennan [1] used analytical techniques with different experimental performances to study interfaces in long fibers reinforced composites. These methods were: Energy Dispersive X-Ray Spectrometry (EDX), and Scanning Auger Microprobe (SAM).

We fixed an experimental procedure to study both the interfaces and the diffusion or inter-diffusion processes in long fibers reinforced composites. On one hand, we used secondary ion mass spectrometry (SIMS) and, on the other hand, EDX and Electron Energy Loss Spectroscopy (EELS). SIMS has a relatively poor lateral resolution ($\sim$ 1.4 $\pm$ 0.5 μm) but an excellent resolution in depth (a few nm) and a high sensitivity (10^{-4} at. % for oxygen). EELS has a relatively poor sensitivity ($\sim$ 0.5 at.%) but an excellent lateral resolution ($\sim$ 1 nm); EELS experiments were made in transmission mode and spectra were sensitive to specimen thickness (50 to 60 nm).

Initially, we studied by SIMS fibers coated by chemical vapor deposition (CVD): Nicalon SiC fiber 202 / C-CVD / SiC-CVD prepared by ONERA-Chatillon*. The carbon layer was introduced to control the chemical bonding between the fiber and the SiC coating. The carbon layer ($\sim$ 100 nm) was of particular interest for checking the ability of SIMS in interfacial layer analyses.

Then, we compared SIMS and EELS analyses of the same coated fiber element [2]. Despite the heterogeneity of the materials considered (SiC fibers as well as SiC-CVD), reasonable agreement was observed between the results obtained by both techniques.

Finally, we studied a fiber / glass-ceramic matrix composite prepared by ONERA. We used (i) SIMS to follow the variation of composition through out the composite and (ii) EDX or EELS to characterize segregations or inclusions in the interfacial layer or in the matrix. The first results are described and compared to the ones observed by Brennan.

*Office National d'Etudes et de Recherche Aerospatiales - Division Matériaux.

EXPERIMENTAL TECHNIQUES:

EELS and EDX analyses:

Analyses were made using a VG HB 501 transmission electron microscope fitted with a Gatan 607 spectrometer and with a standard Si-Li detector. They were run in a fixed probe mode. The analysed volume is a function of both the probe size (1 nm) and the sample thickness; It was roughly equal to 60 nm^3. From the recorded spectra, the elements were identified and their relative atomic concentration calculated [3].

Coated fibers were prepared by ion thinning. For fibers / glass composite (1D), transverse cross sections were obtained by mechanical polishing and ion thinning.

SIMS:

Coated fibers were fixed on an aluminium holder.

The best sensitivity was reached for Si, C and O when primary Cs^+ ions were used and negative secondary ions analysed. The collection set up was adjusted so that the collected secondary ions were emitted by a 3 μm^2 surface.

Each fiber was eroded radially from its surface to its center, and the variation of the secondary ionic intensities Ai^n was measured as a function of the erosion time t (crater mode). Knowing the erosion rate of the fibers and of the deposits, the profiles $Ai^n = f(depth)$ were deduced from $Ai^n = f(t)$. On the profiles shown, the successive data points correspond to slices of material which are distant of 10 nm and whose volume is $\sim 10^7$ nm^3.

The composite samples were cut so that fibers lay parallel to the surface.

The best sensitivity for Mg, Zr, Al and Li was obtained when both a primary O_2^+ ion beam was used and analysis of the positive secondary ions was made. The experimental conditions (erosion, collection), were the same as for the coated fibers.

Analyses were made by maintaining the sample fixed under the incident beam (crater mode) or by moving the sample under the beam along a direction perpendicular to the fibers, to obtained a surface analysis (scanning mode).

In all experiments, the ratios Ai^n / Aj^m were calculated to eliminate the influence of possible fluctuations in the experimental conditions (erosion,collection). The atomic ratios were given by $(Ai)/(Aj) = k \ Ai^n / Aj^m$ and the calibration factors k were obtained from the analysis of standards which have a composition nearly equal to $(Ai)/(Aj)$.

MATERIALS:

Coated Nicalon SiC fibers (202):

SiC deposits were made on fibers segments, whose length were equal to 4 cm, under the conditions given in table 1 [4]. Their thickness in the median part of the fibers is also given in table 1. The deposits consist of two layers (Fig. 1): the first of which - so called CVD1- is constituted of nanometric grains, the second one - CVD2- is made of micrometric grains [5].

Nicalon SiC fibers (202) / glass - ceramic matrix composite:

The glass contains Al_2O_3, Li_2O, MgO and ZrO_2 as main additives. The processing of the composite is described in ref. [6]. Scanning or Transmission Electron Microscopy (SEM or TEM) show that the composite consists of fibers, whose diameters vary from 8 to 20 μm,

Table 1: Thickness of the CVD layers; Carbon layers were deposited at 1273 K , under 133 Pa during 15 mn and SiC deposits at 1573 K, under 66 Pa , during 30 mn.

Samples	C (nm)	SiC – CVD1 (µm)	SiC – CVD2 (µm)
F1	65	2.3	2.7
F2	∿ 100	0.5	2.7
F3	∿ 100	2.5	3.0

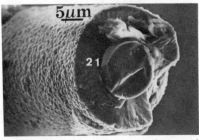

Figure 1:Scanning electron micrograph of the coated fiber F2. It shows the CVD1 and CVD2 layers.

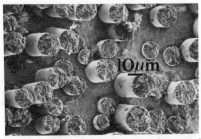

Figure 2:Scanning electron micrograph of the fibers/glass composite etched with HF acid.

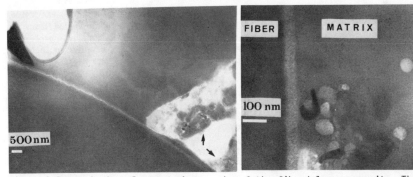

Figure 3:Transmission electron micrographs of the fiber/glass composite. They show the glass matrix partially ceramized (arrows), inclusions and the interfacial layer.

Table 2: Atomic ratios C/Si and O/Si of coated fibers determined by EELS or by SIMS .

Sample	Method	C / Si			O / Si		
		fiber	CVD1	CVD2	fiber	CVD1	CVD2
F1	EELS	1.6 ± 0.2	3.0 ± 0.6	–	0.6 ± 0.2	–	–
	SIMS	1.6 ± 0.2	1.3 ± 0.2	0.9 ± 0.2	0.4 ± 0.2	$(4 ± 2) 10^{-4}$	$(2 ± 1) 10^{-4}$
F2	SIMS	1.4 ± 0.2	0.9 ± 0.2	0.8 ± 0.2	0.4 ± 0.1	$∿ 2\ 10^{-3}$	$(1 ± 1) 10^{-4}$
F3	SIMS	1.5 ± 0.1	1.3 ± 0.2	1.0 ± 0.2	0.4 ± 0.1	$(4 ± 1) 10^{-4}$	$(4 ± 1) 10^{-4}$

split in a partially ceramized glass matrix (Fig. 2 & 3). The matrix contains light or dark inclusions, whose dimensions and sizes are variable. A light layer of 50 to 100 nm is observed at the fiber / matrix interface. Strains occuring during the sample preparation produced sometimes cracks either inside the layer or at the layer / matrix interface.

RESULTS AND DISCUSSION:

Sims analyses of coated SiC fibers:

In the median part of three coated fibers segments, F1, F2 and F3, respectively 4, 5, and 2 quantitatives analyses were performed. Figure 4a represents the variation of the secondary ionic intensities C^-, Si^-, and O^- during an analysis of F1. Figure 4b shows the corresponding C^-/Si^- and O^-/Si^- ratios.

In any given analysis, ratios were known within a few per cent. The mean values in F1, F2 and F3 are given in table 2. The C^-/Si^- ratios in the CVD1 and the O^-/Si^- ratios in the fibers were measured on the plateau exhibited respectively by the curves $C^-/Si^- = f(x)$ and $O^-/Si^- = f(x)$. The spread between experimental values - 10% to 20% - attest the heterogeneity of the materials. Accuracy on the atomic ratios which include errors on the calibration coefficient k, vary from 20% to 30%.

From the results given in tables 1 and 2, the following conclusions can be drawn:

(i): The characteristics of the CVD deposits vary greatly at the beginning of the reaction (CVD1) and become reproducible (CVD2).

(ii): The atomic ratio C/Si is roughly constant in the coated fibers and its mean value is equal to 1.5 ± 0.3. It is slightly higher than the ones obtained in uncoated fibers by SIMS: C/Si = 1.1 ± 0.4 [10]. Analyses by others methods gave C/Si = 1.6 (electron probe [7]) or C/Si $\sim$ 1.4 (SAM [1]). New analyses are planed to establish whether or not the coated fibers exhibit higher carbon contents than the uncoated fibers used.

(iii): In the bulk, oxygen contents are similar in coated or uncoated fibers: O/Si $\sim$ 0.4 in coated fibers; In uncoated fibers O/Si = 0.4 ± 0.1 (SIMS [10]), O/Si $\sim$ 0.3 (electron probe [7]), or O/Si $\sim$ 0.2 (SAM [1]). Near the surface, uncoated fibers are richer in oxygen than in the bulk (SAM [1]) wheareas coated fibers are less rich in oxygen up to a depth of $\sim$ 1 μm (fig. 4b). Therefore oxygen diffuses out of the fiber during the processing.

(iiii): Fibers consist of SiC and C - as shown from electronic diffraction patterns - and probably also of SiO_2 [8]. In the coated fibers, the molar concentrations are: SiC $\sim$ 47%, C $\sim$ 41%, SiO_2 $\sim$ 12%.

Studying the O^-/Si^- and C^-/Si^- profiles, we tested the ability of SIMS technique to analyse interfacial layers. The following results were obtained:

- Qualitative analysis of interfacial layers is possible at least down to a thickness $\Delta x \sim$ 65 nm because the carbon layer was always identified in our experiments.

- Quantitative analysis is impossible in an interfacial layer of Δx = 200 - 400 nm:
In the experiment shown in figure 4, the variations of composition occur on $\Delta x \sim$ 200 nm both on the O^-/Si^- profile at the interface CVD1/CVD2 and on the C^-/Si^- profile at the interface CVD1/C or C/fiber.
This apparent interface thickness is not only due to diffusion as observed on both profiles; rather it is due to the decrease of the gradients at the interface resulting of the three following factors: (i) the distance between successive slices of materials in which one element is analyzed, (ii) the curvature of the interface, (iii) the rugosity at the end of the crater due to the variation of the erosion rate as a function of the chemical composition or of the grains orientation.
In all the experiments Δx varies from 200 nm to 400 nm.

EDX and EELS analyses of a coated fiber:

Qualitatives analyses of F1 were performed by EDX and by EELS. Both techniques show that F1 contains only Si, C, O at concentrations higher than ⌄ 0.5%.

Several quantitative EELS analyses were carried out on the sample: 4 in the fiber, at a distance from the interface ranging from 2 to 250 nm in areas free from inclusions, 2 in the fiber inclusions and 4 in the CVD1 deposit at 2 or 200 nm from the carbon layer (the CVD2 was fully removed during the ion thinning).

Typical EELS spectra are shown in figure 5. Analytical results are given in table 2. The dispersions between the measurements are greater than the experimental accuracies; They are not related to the distance between the analysed areas and the interface; They show the heterogeneity of the materials. The oxygen content in the fiber is close to the sensitivity limit; it is the reason why oxygen was not detected in one of the four spectra.

In the fiber inclusions, the carbon content and the oxygen content (C/Si ⌄ 8 and O/Si ⌄ 4) are clearly greater than in the surrounding matrix (C/Si = 1.5 and O/Si = 0.4). Electronic diffraction patterns show that the inclusions contain free carbon. New experiments have to be performed to determine the chemical bonding of oxygen in the inclusions. Inclusions were previously identified by Sawyer [9]. Their density in the fiber seems to be less than one per cent in volume.

Taking into account the uncertainties on the cross sections, atomic ratios measured in these experiments are known within 20% (C/Si in the fiber) to 40% (O/Si in the fiber).

When analysing SiC, the EELS spectra exhibit a plasmon peak at 21.5 eV. In CVD1, such spectrum was found at 200 nm from the interface (Fig. 6).But at ⌄ 2 nm from the interface, a change was observed on the plasmon peaks: peaks at 16 eV and 24.5 eV were clearly identified (Fig; 7). These plasmons are respectively characteristic of Si and C. Therefore, it seems likely that aggregats of Si and C exist in the analysed volume and its immediate surrounding. These illustrate the heterogeneity of the CVD1.

Comparison between EELS and SIMS analyses:

When C/Si was measured in the fibers, the large carbon inclusions were taken into account only by SIMS,thus creating a difference of only a few per cent between the two approaches.

Superficial contamination of samples often alters carbon content measurements in EELS analyses. During our experiments, neither spots due to contamination, nor changes in the carbon peak as a function of time were observed. Therefore, either the amount of hydrocarbons on the specimen surfaces was extremely low, or these were unaffected by irradiation. In the fiber, the excellent agreement between EELS and SIMS measurements shows that the possible influence of contamination is smaller than the spread of experimental values. In CVD1, the disagreement between EELS and SIMS may be due to local contamination; However, it may also confirm the heterogeneity in the CVD1 layer previously mentionned.

The oxygen contents measured by EELS and SIMS are equal within the uncertainties. This agreement is particularly significant as measurements of oxygen contents by EELS are independent of any contamination.

Agreement observed between EELS and SIMS measurements in fibers demonstrates their reliability and shows that both techniques can be used to reach a complete characterisation of fibers reinforced composites.

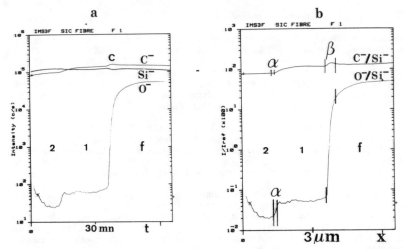

Figure 4:SIMS profiles. Secondary ionic intensities C^-, Si^-, O^- versus time (a), C^-/Si^- and O^-/Si^- ratios versus depth (b) respectively in the fiber (f), in the CVD1 (1), in the CVD2 (2) and in the C deposit (C). Slashes delimit zones of 200 nm or of 500 nm (β). The thickness of the C deposit was equal to 65 nm.

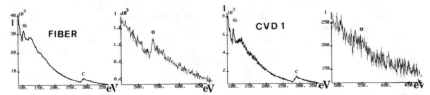

Figure 5: EELS spectra obtained in the coated fiber F1.

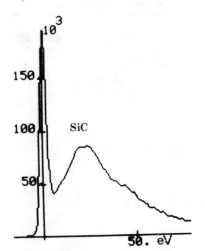

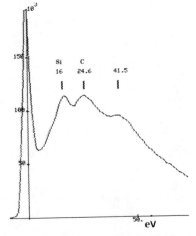

Figure 6: Plasmon peak obtained at ∿ 200 nm from the interface C/CVD1 in the CVD1 deposit.

Figure 7: Plasmon peaks obtained at ∿ 2 nm from the interface C/CVD1 in the CVD1 deposit.

APPLICATION OF SIMS AND EDX - EELS ANALYSES TO A LONG FIBER REINFORCED COMPOSITE:

In a material as complex as the fibers/glass composite, quantitative analyses by SIMS would need a lot of standards. As a first approach, a qualitative analysis using the scanning mode was done. The variation of Si, C, Al, Li, Mg and Zr amounts were checked in the composites. SIMS profiles are shown in figure 8. Due to the diameter of the area analysed ($\emptyset \sim 2\ \mu m$), the decrease of concentration gradients occurs on $2\ \mu m$ on either side of the interfaces I1, I2 and I3. Interfaces have been located at the inflexion points of the Si profiles. This criterion was choosen because the amount of Si diffusion from fibers into the matrix is likely restricted.

This investigation was completed by a few EDX and EELS analyses. Mg and Zr contents are within or below the sensitivity of these techniques. Atomic concentrations determined by EDX are given in table 3.

From SIMS profiles, EDX and EELS analyses, the following initial conclusions were reached:

- **Fibers:** Mg, Al and Li were not detected by EDX or EELS in the fiber. SIMS profiles prove the diffusion of Mg and Al into the fiber up to a depth of several microns; Li also probably migrates into the fiber.
The penetration of Zr into the fiber, which is likely detected on SIMS profiles, has been confirmed by EDX.

- **Matrix:** The amounts of Al and Zr vary in the matrix as well as in the inclusions (cf. Fig. 8 and table 3). Black inclusions are richer in Zr than the light ones and some consist of Zircon. Zircon inclusions were also found by Brennan in comparable SiC fibers / LAS matrix composite.
If carbon diffuses from the SiC fibers into the matrix, the penetration depth seems to be less than 4 microns (Fig. 8).

- **Interfacial layer:**
The interfacial layer is richer in Al and Zr than the fibers but poorer than the surrounding matrix.
Areas rich in carbon were found by EELS in light inclusions close to the interfacial layer, in the interfacial layer and at the layer/fiber interface. Carbon does not seem to be the unique constituant of the light interfacial layer as found by Brennan in his own material. In some places at least, the layer contains noticeable amounts of Si, O and sometimes Al, Zr and Li. EELS spectra show that Si and O present in the interfacial layer form SiO_2 molecules (Si edge = 106 eV); This would mean that the interfacial layer develops in the matrix on contact with the fiber.

CONCUSIONS:

SIMS analyses of Nicalon SiC fibers, coated with C and SiC by CVD have been performed. They show that the CVD process develops in two steps: initially, the composition of the deposit (CVD1) varies as a function of time and from one experiment to the other ; subsequently, the composition of the deposit is reproducible (CVD2) and roughly stoechiometric. The SiC fibers contain large amounts of free C ($\sim 41\%$) and of SiO_2 ($\sim 12\%$).

SIMS proves its capability to perform quantitative analyses of fibers reinforced composites through out the material, exept near the interface. Qualitative analyses on the interfacial layer are possible, at least when the thickness is higher than 65 nm.

Good agreement was observed among the coated fiber compositions as measured by SIMS and EELS . It shows that both techniques can be used to reach a complete characterisation of fibers reinforced composites.

A first characterization of Nicalon SiC fibers / glass matrix composite is in reasonable agreement with Brennan previous study on a similar material.

238

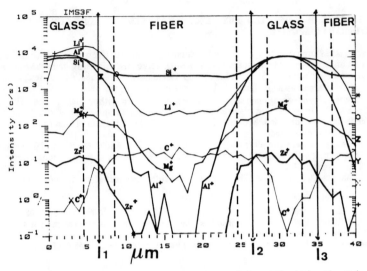

Figure 8: SIMS profiles. Secondary ionic intensities Si+, Li+, C+, Mg+, Al+ and Zr+ obtained by the scanning mode in the fibers/glass composite.

Table 3: Atomic composition of fiber/glass composite determined by EDX.

element	fiber	interfacial layer	matrix	inclusions dark		light
Si	99.4	90 ± 8	89	47 ± 2	26 ± 6	73 ± 3
Al	-	8 ± 8	7	7 ± 3	6 ± 1	14 ± 3
Zr	0.6	2 ± 2	4	46 ± 2	68 ± 7	13 ± 2

REFERENCES

1. J.J. Brennan, Proc. Conf. Tailoring Multiphase & Composites Ceramics, Penn. State Univ.,17-19 July 1985.

2. F. Anxionnaz, M. Schuhmacher, P. Trebbia, M. Lancin, J. Microsc. Spectr. Electronique, April 1987.

3. C. Colliex, Adv. in Optical & Electron Microsc., 9, p. 65 (1984).

4. The material was provided by L. Grateau & M. Parlier, ONERA Chatillon France;
   M. Parlier, Thèse Ecole Sup. Mines de Paris, 1984.

5. F. Anxionnaz, M. Lancin, unpublished results.

6. The glass was provided by Saint - Gobain France, the composite by J. Jamet. J. Jamet, M. H. Ritti,
   P. Peres, ONERA Report N° 163548; J. Jamet Carrefour Int. Ceramique, Limoges, 6 _ 10 Octobre 1986.

7. Leniewsky, DRET Report N° 84445, 1986.

8. P. Simon, Thèse Ecole Sup. Mines de Paris, 1984.

9. Sawyer, R.S. Aron, F. Haimbach, M. Jaffe, K.D. Rappaport, in Ceramic Engeneering and Science, Ed. Am.
   Ceram. Soc., Proc. 9th Annual Conf. on Composites and Advanced Materials, Vol. 6 N° 7-8, Aug. 1985.

10. M. Schuhmacher, J. S. Bour., M. Lancin to be published.

PART V

Fracture and Deformation
Behavior in Ceramic Composites

STRAIN AND FRACTURE IN WHISKER REINFORCED CERAMICS [1]

PETER ANGELINI,[*] W. MADER,[**] and P. F. BECHER[*]
[*] Oak Ridge National Laboratory, P.O. Box X, Oak Ridge, TN 37831
[**] Max-Plank-Institut für Metallforschung, Stuttgart, Federal Republic
of Germany

ABSTRACT

Whisker reinforced ceramics offer the potential for increased fracture
strength and toughness [2]. However, residual strain due to the thermal
expansion mismatch between Al_2O_3 and SiC may affect mechanical properties of
such composites. Crack tip interaction with the whisker/matrix may lead to
changes in debonding behavior or influence other toughening mechanisms. The
strain field in the Al_2O_3 matrix surrounding SiC whiskers was analyzed with
a High Voltage Transmission Electron Microscope (HVEM). Strain contrast
oscillations indicating the presence of residual stress in the specimen
were observed in a Al_2O_3-5 vol % SiC composite having ≈15 μm grain size.
The strain field was found to have both radial (perpendicular to whisker
axis) and axial (parallel to whisker axis) components. A strain field
was also present near the end faces of SiC whiskers. In situ thermal
annealing to 573, 873, and 1173 K showed a decrease in the residual strain
while in situ cooling to ≈77 K revealed little change in the strain. These
results show that residual stresses in the compacts result from differences
in thermal expansion and elastic constants of the matrix and whisker
materials. Dynamic in situ fracture experiments performed in an HVEM on the
Al_2O_3-5 vol % SiC material having ≈15 μm as well as on Al_2O_3-20 vol % SiC
having ≈1 μm grain size revealed that fracture resistance is due to a number
of mechanisms including debonding near the whisker matrix interface, crack
deflection, pinning, and bridging by SiC whiskers. Formation of secondary
fractures and microcracks near and in front of propogating crack tips was
also observed in the larger grain size composite.

INTRODUCTION

Previous results and analyses of fracture surfaces which were either
perpendicular or parallel to the hot pressing axis (HPA) of Al_2O_3-SiC
composites reveal the SiC whiskers to be distributed anisotropically [3].
The whiskers tend to be oriented with their longitudinal axis in planes
perpendicular to the HPA. However, they are randomly oriented within those
planes as shown in Fig. 1. This type of whisker orientation can influence
the contribution which any particular toughening mechanism offers toward
increasing fracture toughness. The fracture surfaces of bulk specimens are
irregular with debonding occurring near the matrix-whisker interface. Both
intergranular and transgranular fracture of matrix grains are observed. The
short segments of whiskers with longitudinal axis nearly normal to the surface are also noted as evidence of a minimal amount of whisker pullout.
Such short whisker segments imply that stresses larger than the fracture
strength of whiskers were reached during crack propagation.
Identification of residual thermal stress present near SiC whiskers
is important in order to determine its influence on mechanical properties.
Residual thermal stresses can develop both in the matrix and whiskers during
cooling of a composite from its fabrication temperature. The stresses
result from differences between a number of properties of the matrix and
whisker materials. The residual stress [4], at the interface of a spherical
particle located in a matrix, upon cooling is represented by the following
Eq. 1,

Mat. Res. Soc. Symp. Proc. Vol. 78. ⃝c 1987 Materials Research Society

$$P = \frac{\Delta\alpha\Delta T}{\dfrac{1+\nu_M}{2E_M} + \dfrac{1-2\nu_p}{E_p}}$$ Eq. (1)

where P = stress at the interface, $\Delta\alpha$ = thermal expansion difference between the matrix and precipitate material, ΔT = temperature difference, ν_i = Poisson's ratio for material i, and E_i = Young's modulus of the respective component i.

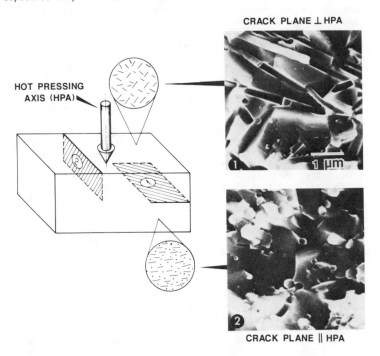

CRACK PLANE $\perp$ HPA

HOT PRESSING AXIS (HPA)

CRACK PLANE $\parallel$ HPA

Fig. 1. Whiskers tend to be oriented with their axis perpendicular to the HPA, however, within those planes are randomly oriented.

Although the residual thermal stress at or near the interface of the matrix and whisker would not have the exact relationship shown in Eq. 1, it is expected to have similar form. Residual stress is related to the Poisson's ratio and elastic modulus of the whisker and matrix materials, should decrease as ΔT decreases and be proportional to the difference in thermal expansion between Al_2O_3 and SiC. The linear thermal expansion coefficients for Al_2O_3 or SiC do not change significantly above ambient temperature, however, their magnitude and difference become much smaller at temperatures below ambient [5], [6]. Thus the shape of the residual stress versus temperature curve is expected to have a maximum value at low temperature, remain near the maximum value up to near ambient, and decrease to possibly zero near the specimen fabrication temperature. Since the thermal expansion coefficients of Al_2O_3 and SiC are $\approx 8 \times 10^{-6}$ and 4×10^{-6} respectively at ambient temperature, the total misfit between Al_2O_3 and SiC, for a temperature interval between 2123°K and 300°K, is approximately 0.7%. The misfit translates to less than 4 nm for a whisker of 0.5 μm diameter. Quite small misfits are thus expected to occur in this system and the presence of amorphous or crystalline boundary layers, microcracks, or variations in whisker diameter along its axis will affect the stress distribution. Chemical reactions between the matrix and whisker can also lead to boundary layers and possible chemical bonding. If chemical bonding were to occur then toughening mechanisms such as debonding or bridging may not occur to the same extent. Effect of impurities [7-9] or sintering aids on matrix-matrix grain boundaries and matrix-whisker interfaces can also affect high temperature mechanical and chemical stability as well as residual stress in the composite.

Stress levels in composites can be measured by Transmission Electron Microscopy (TEM). A technique for determining the misfit strain of inclusions whose dimensions are smaller than the extinction distance of a diffracting vector used for imaging has been developed [10], [11]. The model was developed for spherical precipitates having the following radial displacement field [12], μ,

$$u = \frac{\varepsilon r_0^3}{r^2} \text{ for } r > r_0 \qquad \text{Eq. (2)}$$

and included only the elastic component to the displacement field. The TEM image width, (IW), or distance from the precipitate to a point in the matrix where the image intensity is at a defined fraction of the background level, was shown to have the following form [10], [11]

$$IW \propto \log_{10}\left(\frac{\varepsilon g r_0^3}{\xi_g^2}\right) \qquad \text{Eq. (3)}$$

where ε = elastic strain component at $r = r_0$, r_0 = radius of the spherical inclusion, ξ_g = extinction distance for the operating diffraction vector, g. Two experimental determinations of strain near amorphous silica precipitates in an internally oxidized Cu-Si alloy resulted in measured ε values of 6×10^{-3} and 1×10^{-3} depending upon the quenching rate from 1073°K. The value of 6×10^{-3} for the fast quench compared favorably with a calculated ε of approximately 9×10^{-3}. Measurements near γ-Al_2O_3 precipitates in an internally oxidized Cu-Al alloy yielded an ε of approximately one half of that calculated. The above relationship points out some of the important parameters affecting these types of measurements as performed in a TEM and offers guidance for the experimental observations in the present work. Decreasing the extinction distance tends to increase the distance of an effective "fringe" from the inclusion. Since it is difficult at 1 MV to obtain two beam diffraction conditions, the effective ξ_g is nearly always smaller than that for the two beam case. These variations can be calculated, but actual corrections depend on the intensity of each diffracting g vector. The excitation error, s, is important in determining the position of the "fringes" from the inclusion. Experimentally it is most convenient to make all measurements with s = 0 (exact Bragg condition for the strongest operating diffraction vector). Eq. (3) also points to a strong dependence on the radius of the inclusion. In the present case, whiskers are generally cylindrical but with nearly hexagonal cross section. Thus errors could be introduced in future formulations with cylindrical approximations due to the use of an effective radius r_0'.

The strain field surrounding nearly spherical ZrO_2 inclusions in Al_2O_3 matrix grains has also been observed by TEM [13], [14]. The type of contrast noted in those studies was similar to that observed in the present research and is useful when the extinction distance of an operating diffraction vector is approximately equal to or smaller than the diameter of the inclusion. An explanation of the contrast oscillations has been given [15], [16] for the Al_2O_3-ZrO_2 case. Calculated images were compared with experimental ones and the misfit strain parameters between ZrO_2 and Al_2O_3 obtained as a result of image matching. In that work the measured misfit was approximately one half of the calculated value. This indicates the possible occurrence of (1) stress relaxation occurring due to high fabrication temperatures of typically 1700 K, (2) relaxation due to thin foil effects, or (3) presence of microcracks or boundary layers.

Prior to the present work, TEM observation of strain field oscillations in the matrix surrounding whiskers had not been performed. Experimental results of this are presented in this paper. Models based on a cylindrical geometry and image matching techniques are presently being developed.

Dynamic in situ fracture experiments on SiC whisker reinforced Al_2O_3 were performed in an HVEM to identify fracture toughness mechanisms operating in these composites and to note the interaction of crack tips with whiskers. The state of stress in thin specimens such as those used for in situ fracture is complicated. Although the overall specimen is strained in tension the stress field in the electron transparent regions is not in uniaxial tension and various aspects such as stress concentration near voids or crack tips must be addressed. A typical in situ fracture specimen prior to being loaded onto the holder is shown in Fig. 2. The electron transparent "ridges" are the two small segments near the center of the specimen and between the three nearly circular holes introduced during ion milling. Stress is concentrated near the edges of the "ridges" due to the presence of the circular holes. The stress field for a simplified case, where a uniform tensile stress is applied at a large distance from an elliptical hole, located in the center of a thin plate of uniform thickness [17-20], points to some important considerations. The elliptical hole is assumed to have a major axis of length, 2 a, oriented perpendicular to the applied stress, and a minor axis of length 2 b. The maximum stress, σ_y, occurs at a point, A, on the edge of the cavity and located on the centerline of the specimen. The maximum stress, σ_y, is in the direction of the applied stress and is given by

$$\sigma_y = \sigma[1 + 2 \ a/b] \ . \qquad \qquad \text{Eq. (4)}$$

The stress rapidly decreases with distance from the edge of the cavity and becomes nearly equal to the applied stress within a distance equal to a few radii of curvature from the edge of the cavity. There is also a stress, σ_x, acting in the direction normal to the applied stress. The stress, σ_x, is zero at point A, reaches a maximum value at a point between the edge of the hole and the edge of the specimen located at $\approx b^2/a$ from the edge of the hole, and then decreases slowly to zero. This indicates the specimen is not in a uniaxial stress condition. This analysis can also be extended to a crack tip [21].

Shear stresses also exist near cavities or crack tips. A maximum shear stress of greater than six times the applied stress can occur within a small volume near the crack tip [20]. It is, however, reduced to approximately 1.5 σ at a distance less than b^2/a from the crack tip. The effect of crack tip radius decreases rapidly so that the stress distribution near the crack tip is similar irrespective of crack tip radius of curvature. Similar shear stresses also exist near cavities. The largest shear stress occurs near point A, which for the plane stress case (thin specimen), is $\sigma(1 + 2 \ a/b)/2$ on a plane 45 degrees to the x and y axes. The overall ratio of the maximum tensile to maximum shear stress can also be important in determining whether fracture will occur by cleavage or theoretical shear stress. Failure due to cleavage usually occurs when $\sigma_{max} \approx \tau_{max}$. Thus even whiskers whose axes are misoriented near a crack tip can also deflect a fracture or bridge a crack especially if the whisker-matrix interface fails in shear.

The behavior of crack growth may be similar between in situ or thin (plane stress) specimens, and bulk or thick (plane strain) specimens. From Griffith's theory of fracture [22], for an elastic solid under plane strain (thick plate) subject to a uniform stress, the total free energy change is the sum of the elastic strain energy and the surface energy of the crack. The criterion that notch growth occurs (fracture propogates) for an elliptical cavity with a major axis of length 2 a, oriented perpendicular to the applied stress, and minor axis of length 2 b in a thin specimen (plane stress) is

$$\sigma = \left(\frac{2 \ E\gamma_s}{\pi a}\right)^{1/2} \qquad \text{Eq. (6)}$$

and for a thick specimen (plane strain) is

$$\sigma = \left(\frac{2 \ E\gamma_s}{(1-\nu^2)\pi a}\right)^{1/2} \qquad \text{Eq. (7)}$$

where, γ_s = surface energy, E = Young's modulus, and ν = Poisson's ratio, and a>>b. The difference between the plane stress (thin specimen) versus the plane strain (thick, or bulk, specimen) is the factor $1/(1-\nu^2)$ which for a Poisson's ratio of 0.3, is ≈1.1. This is not a large difference, however, if there is a nonlinear zone around the crack then plane stress versus plane strain results would be quite different. It is also important to note that since the stress needed to produce fracture is also proportional to the surface energy to the one half power, processes which tend to decrease γ_s can be very effective in enhancing crack growth. Surface energy, γ_s, of Al_2O_3 also varies depending upon the type of impurity and content, and type of crystallographic plane acting as the surface. If cleavage occurs then the orientation of grains with respect to one another is also important as the fracture energy of single crystal Al_2O_3 [23-25] for the planes (11$\bar{2}$0), (1$\bar{1}$00), and (0001) are 6.0 7.3 and >40 in units of J/m^2 respectively.

The stress distribution for fibers embedded in a matrix and subject to a tensile stress parallel to their axes shows that shear stresses exist at the fiber-matrix interface [26]. However, in that study (1) bonding of the matrix and fiber at the ends of the fiber is neglected so that no load is transferred across the ends, (2) stress concentration near the end of the fibers is not included, and (3) residual stresses due to thermal misfit were also neglected. This type of whisker orientation relative to applied stress is the most favorable for improving fracture toughness by a pullout mechanism. As the misorientation of fibers increases, the contribution to toughening can decrease. The largest interface shear stresses occur near the ends of fibers and decrease to zero at the middle position along the fiber's length. The analysis leads to a critical length which must be less than one half the length of fibers to minimize fiber fracture at the crack plane thereby obtaining the greatest toughening and strengthening effect. Thus thermal expansion as well as mechanical property differences between the matrix and whisker material are important in determining stress fields and their subsequent effects in composites. Additional factors such as mechanical binding due to variations in diameter of whiskers along their length, chemical bonding, boundary layers, and impurities at boundaries can also influence the stress distribution and mechanical properties of composites.

In situ fracture experiments have previously been performed on ZrO_2 transformation toughened ceramic materials [27]. In those studies areas of specimens were analyzed prior to and after a crack was propogated through the region. In this way the transformation zone in front and to the side

246

of the fracture was measured and related to the number and size of ZrO_2
particles which had transformed to monoclinic symmetry. Thin foil effects
were shown to be minimal when ZrO_2 inclusions were completely confined
within the matrix. The in situ straining holder used in the present work
was the same as that used in those previous studies. Similar in situ before
and after crack propogation studies were performed. However, the present
study also contains dynamic observations of stationary as well as moving
crack tips while the specimen was under stress.

EXPERIMENTAL

One of the composites evaluated contained 20 vol % SiC whiskers in an
Al_2O_3 matrix. The compact also contained 0.5 wt % MgO and 2.0 wt % Y_2O_3 as
sintering aids. A second compact with 5 vol % SiC in an Al_2O_3 matrix and
no sintering additives was also studied. The materials were hot pressed at
2123°K in a carbon die [2], [3]. The average grain size of the 20 vol % SiC
compact was ≈1 μm as compared to a grain size of ≈15 μm for the 5 vol % SiC
specimen. One of the important factors which must be met for obtaining
dense specimens is the dispersal of SiC whiskers. This can sometimes be
difficult since whiskers tend to agglomorate as shown in Fig. 3. Specimens
for the strain contrast experiments were obtained by mechanically cutting,
grinding, polishing to a thickness of ≈60 μm, and then Ar ion milling at
6 kV to electron transparency.

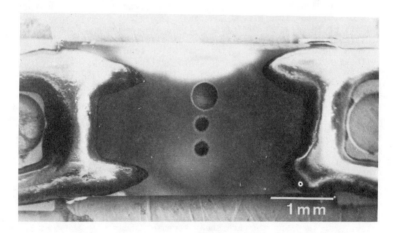

Fig. 2. SEM micrograph of in situ straining specimen.

The specimen preparation procedure for thinning in situ straining
specimens was more involved. The specimens were cut, ground and polished
to a rectangular shape of 3 mm width, 5 mm length, and 50 μm thickness.
Two dimples approximately 37 μm deep were placed near the center and then
the specimen was ion milled with Ar at a voltage of 6 kV. Finally it was
mounted onto two copper end support pieces with epoxy. A scanning electron
microscopy (SEM) micrograph of one specimen is shown in Fig. 2. The procedure generally results in a specimen having two electron transparent "ridges"
between three nearly circular holes. The thickness of the ridges is not
constant but increases from nil at the edge to approximately 5 μm at the
thickest section near the center.

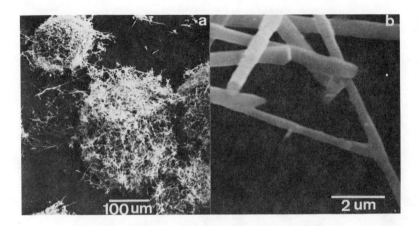

Fig. 3. As received SiC whiskers.

The transmission electron microscopy was performed with an AEI-EM7 High Voltage Electron Microscope operated at 1 MV. In situ heating and cooling was performed with two separate Hitachi double tilt holders. A double tilt straining stage was used for in situ fracture experiments [28], [29]. The straining rate, without a specimen in the holder, could be varied from ≈100 nm/sec to 1 μm/sec; however, slower straining rates to ≈15 nm/sec were possible. The in situ fractures were recorded at real time using a television and video tape recording system present on the microscope.

STRAIN CONTRAST RESULTS AND DISCUSSION

Many conditions are important for observing strain contrast oscillations, and several are: (1) constant specimen thickness to remove thickness fringe effects, (2) whisker centered in foil to minimize foil bending and thus minimize bend contours in the image, (3) diffraction vector, g, and electron beam perpendicular to the axis of the whisker to evaluate only the radial component of strain, and (4) depth and diameter of the whisker, extinction distance, and excitation error must be known [30]. Only then does the misfit between the whisker and matrix become a parameter which can be varied when comparing calculated with experimental images.

An alumina specimen with 5 vol % SiC was chosen for the strain contrast experiment to minimize whisker-whisker interactions. Since the grain size of the alumina grains is approximately 15 μm, many whiskers are present intragranularly. The micrographs presented in Fig. 4 show a whisker and alumina grain oriented with diffracting conditions suitable for observing the strain contrast oscillations. Foil bending of the Al_2O_3 matrix grain can lead to different diffracting conditions on either side of the whisker. Tilting the specimen less than one degree can bring the strain contrast oscillations into contrast on the opposite side of the whisker as shown in Fig. 5.

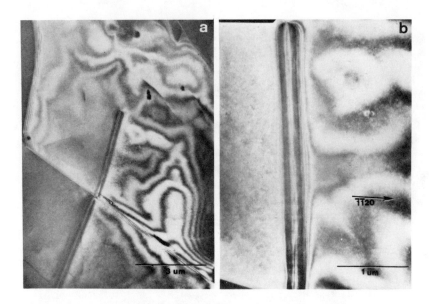

Fig. 4. Strain contrast oscillations present
in Al_2O_3 matrix surrounding SiC whisker.

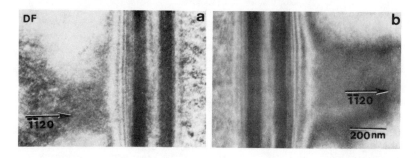

Fig. 5. Strain contrast oscillations in the matrix are
symmetric on either side of whiskers (dark field micrographs).

The micrographs in Fig. 5 are from the region near the center of the
whisker shown in Fig. 4. The position of the contrast oscillations relative
to the whisker are nearly identical on either side of the matrix near the
whisker.

Cooling the specimen in the microscope below ambient is not expected
to significantly change the position of the contrast oscillations since,
if they are due to thermal misfit, the magnitude and difference in thermal
expansion coefficients for both Al_2O_3 and SiC become much smaller in this
temperature range, and the respective ΔT from the fabrication temperature
are not very different. This was found to be the case as can be seen in
the images shown in Fig. 6.

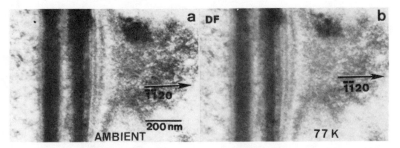

Fig. 6. Lowering the temperature to 77 K did not significantly change the position of the strain contrast oscillations.

Raising the temperature, however, is expected to decrease the residual thermal strain significantly and thus the contrast oscillations should contract toward the whisker. The micrographs presented in Fig. 7 show this behavior. Both the number and distance of the contrast oscillations were found to decrease. Also note the complimentary nature of the bright and dark field micrographs. These images were taken with an excitation error, s, equal to zero as was done for the micrographs shown Fig. 4 to Fig. 7.

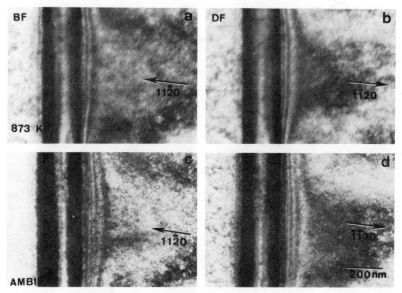

Fig. 7. Strain contrast oscillations; bright field and dark field, right side of the whisker, ambient and 873 K.

Line scans of the intensity of the contrast oscillations versus distance from the surface of the whisker were made from micrographs by using an image analysis system. These data will later be compared with calculated images where the misfit is considered a parameter. However, the position of the last contrast oscillation versus temperature is indicative of the residual thermal stress present in the matrix. The curve presented in Fig. 8 shows the proper behavior as noted previously. The residual thermal stress may not decrease to zero at the fabrication temperature if a residual stress contribution exists due to sintering effects [31].

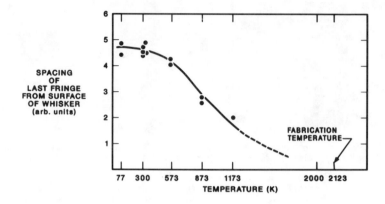

SPACING
OF
LAST FRINGE
FROM SURFACE
OF WHISKER
(arb. units)

Fig. 8. Measured distance of the farthest strain contrast
oscillation versus temperature is related to residual thermal
stress in the matrix.

These data clearly indicate that contrast oscillations are related to
residual thermal stress due to differences in thermal expansion between the
SiC whisker and the Al_2O_3 matrix. Since the diffraction vector chosen for
imaging the contrast oscillations, as in Figs. 1-3, is nearly normal to the
whisker axis, only the radial component of the residual thermal stress field
was evaluated. However, there also exists an axial component to the stress
field as shown by Fig. 9.

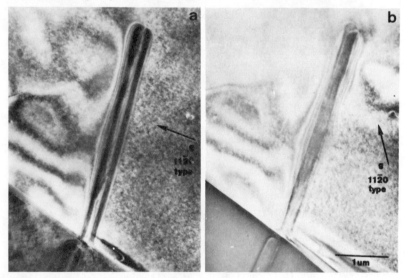

Fig. 9. An axial component to the residual thermal stress
field surrounding SiC whiskers exists in Al_2O_3. Residual stress is
also observed near the ends of whiskers, indicating that stress is
also transferred across the end faces of SiC whiskers.

The analysis of the shear stress at the fiber-matrix interface and of the tensile stress in the fiber mentioned in the introduction did not include the effect of residual thermal stress due to misfit nor did it include stress transfer at the ends of a fiber. Contrast oscillations were also observed with a diffraction vector nearly parallel to the whisker axis (see Fig. 9) which show that residual stresses also exist in the matrix near the end of the whisker. This indicates that stress can be transferred across the ends of whiskers and thus implies that the shear stresses at the whisker-matrix interface may not be as large as expected.

IN SITU FRACTURE RESULTS AND DISCUSSION

In situ fracture experiments were performed in an HVEM to evaluate the mechanisms which occur during crack growth in composites. Straining rates of less than 100 nm/sec were required to keep the propogating crack within the field of view while recording with an on line television system. Specimens of both the 20 vol % SiC and 5 vol % SiC in Al_2O_3 matrix were evaluated. The specimens were sectioned from respective compacts with the plane of the specimen perpendicular to the HPA.

The main objective in one experiment was to note the behavior of material at the crack tip while stress was applied. Bend contours moved during straining and straining was stopped, although the stress was maintained, when the first fractures occurred. The micrographs in Fig. 10 show the subsequent results. One end of the crack can be seen near SiC whiskers Fig. 10(a). Debonding occurred near the whisker matrix interface [Fig. 10(b)]. While obtaining micrographs at high magnification, crack growth continued spontaneously due to beam heating effects. However, the crack was pinned at a SiC whisker [Fig. 10(c)]. Both sides of the fracture near the prior debonded SiC whisker are clearly visible Fig. 10(d). Note that the fracture had restarted at an apex of the hexagonally shaped SiC whisker. This may have resulted from effects of stress concentration effects near the corners. Observations of the crack tip pinned at the SiC whisker did not show any debonding above or below the fracture [Fig. 10(c)]. The radius of curvature of the pinned crack was less than 5 nm as measured from the micrograph. The fracture tip may be approaching an atomically sharp crack as has been noted previously in some ceramics, [32-34]. Continued straining of the specimen showed a completely elastic behavior and dislocation generation or movement was not observed. The fracture continued by a debonding process near the whisker matrix interface until the specimen was completely fractured [Fig. 10(e,f)]. The micrographs in Fig. 11 were taken from the recorded video images during crack propogation. Debonding is clearly visible when comparing Fig. 11(a) and Fig. 11(b).

The fracture surface of this specimen was also observed by SEM analysis after removal from the HVEM. Micrographs were taken along the fracture surface from thick to thin areas and two examples are shown in Fig. 12. The upper micrograph is for a region of approximately 20 µm thickness. The surface is very much like that observed for bulk specimens. Intergranular and transgranular fractures can be seen and debonding is clearly visible. The micrograph of Fig. 12(b) was taken from a region of one of the "ridges" observed during in situ straining in the HVEM. Note the similar appearance of the two fracture surfaces. Similar mechanisms seem to be operating in each case indicating that in situ fractures are closely related to those in the bulk.

A second in situ fracture experiment was performed with a similar specimen (20 vol % $SiC-Al_2O_3$) to observe processes which occur within approximately 10 µm of the crack tip. Debonding, crack pinning, and crack deflection were observed in at least four different dynamic straining experiments. Again no dislocation generation or movement was observed indicating the material fails elastically. Both intergranular as well as transgranular fractures were seen.

252

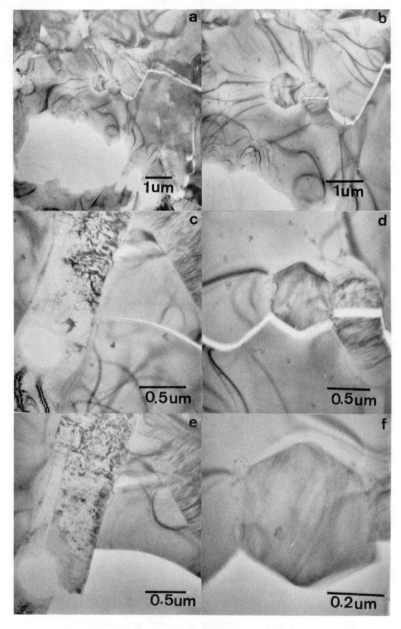

Fig. 10. In situ straining of 20 vol % SiC-Al₂O₃ specimen.

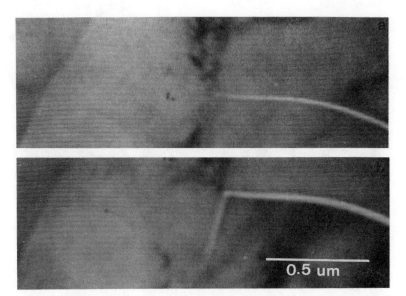

Fig. 11. Debonding occurs during fracture (micrographs taken from video images.

A third specimen was a composite of 5 vol % SiC-Al$_2$O$_3$. Again crack pinning and deflection, and whisker pullout were observed. Debonding occurred near the whisker matrix interfaces. Dynamic recording of stationary and moving crack tips, while the material was being stressed, showed it to deform elastically. An example of the behavior of a crack tip as it approached SiC whiskers and continued beyond them is shown in Fig. 13. This sequence was taken from the video recording and encompasses approximately 10 seconds. The recording however can be reviewed frame by frame. In Fig. 13(a) the crack tip was approaching the SiC whiskers. Fig. 13(b) shows debonding of the whiskers at the crack tip. The fracture continued transgranularly across the Al$_2$O$_3$ grain with strain contours clearly visible as seen in Fig. 13(c). The crack could be made to stop or advance by controlling the strain rate. This particular fracture was stopped and the region in front and to the sides was evaluated. Microcracks were observed within 20 µm in front of the crack tip and secondary fractures were also present. This latter effect is probably related to additional residual stresses introduced from the anistropic expansion properties of Al$_2$O$_3$ versus grain size.

SUMMARY

The strain field in the Al$_2$O$_3$ matrix surrounding SiC whiskers was analyzed in an HVEM. Contrast oscillations were observed which are indicative of strain in the matrix. Very little change was observed in the position of the contrast oscillations upon in situ cooling to ≈77 K. However, for in situ thermal annealing to 573, 873, and 1173 K, both the number of contrast oscillations and their relative distance to the whisker decreased. These results are consistent with the presence of residual thermal stresses due to differences in linear thermal expansion and elastic properties of the Al$_2$O$_3$ matrix and SiC whiskers. Both radial (perpendicular to the whisker axis) and axial (parallel to the whisker axis) components exist to the stress field. Contrast oscillations were also observed at the ends

254

of whiskers indicating stress can be transferred across the end faces of whiskers. In situ straining experiments were performed and fracture mechanisms evaluated. These results showed that fracture resistance is due to a number of mechanisms including debonding near the whisker-matrix interface, crack pinning, bridging, and deflection by whiskers. Formation of secondary fractures near and in front of crack tips was observed in large grained composite. Observation of stationary or moving crack tips while a specimen was under stress revealed that fracture was as elastic process. Dislocation generation or motion was not observed.

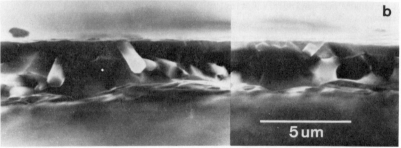

Fig. 12. Fracture surfaces of thick and thin regions for in situ fractured 20 vol % SiC-Al$_2$O$_3$ specimen (SEM images) show similar structures.

ACKNOWLEDGMENT

The authors would like to express their appreciation to A. Strecker for support in maintenance of the double tilt straining holder, D. Waidelich for operational procedures, H. Schadler for maintenance of the HVEM, L. Bitzek for translation of video tapes from the Pal European norm to the NTSC-USA norm, Drs. P. Maziasz and C. Hsueh for reviewing the manuscript, and F. Stooksbury for typing this manuscript.

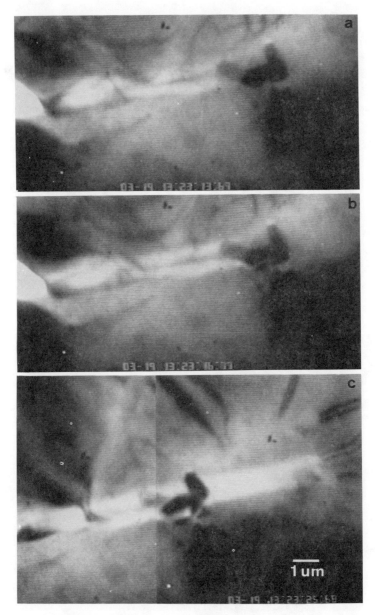

Fig. 13. Movement of crack tip as it approached and continued beyond SiC whiskers. Debonding occurred and deformation was elastic.

256

REFERENCES

1. Research sponsored by the Division of Materials Sciences, U.S. Department of Energy, under Contract DE-AC05-84OR21400 with Martin Marietta Energy Systems, Inc.

2. G. Wei and P. F. Becher, J. Am. Ceram. Soc., 87, 267 (1986).

3. (a) P. F. Becher, T. N. Tiegs, J. E. Ogle, and W. H. Warwick, in Fracture Mechanics of Ceramics, edited by R. C. Bradt, A. G. Evans, D.P.H. Hassleman, and F. F. Lange, Plenum Publ. Corp., New York (1986) pp. 61–64; (b) N. Claussen, R. L. Weisskopf, and M. Rühle, ibid., pp. 75–86.

4. J. Selsing, J. Am. Ceram. Soc., 419 (1961).

5. J. B. Wachtman, Jr., T. G. Scuderi, and G. W. Cleek, Am. Ceram. Soc., 45, 319 (1962).

6. G. A. Slack and S. F. Bartram, J. Appl. Phys., 46, 89 (1975).

7. A. W. Funkenbush and D. W. Smith, Met. Trans., 6A, 2299 (1975).

8. R. S. Jupp, D. F. Stein, and D. W. Smith, J. Mat. Sci., 15, 96 (1980).

9. S. Baik and C. L. White, "Anisotropic Ca Segregation to the Surface of Al_2O_3," J. Am. Ceram. Soc. (1987).

10. M. F. Ashby and L. M. Brown, Phil. Mag., 8, 1983 (1963).

11. M. F. Ashby and L. M. Brown, Phil. Mag., 8, 1964 (1963).

12. N. F. Mott and F.R.N. Nabarro, Proc. Phys. Soc. Lond., 52, 86 (1940).

13. M. Rühle and W. M. Kriven, Ber. Bunsenges. Phys. Chem., 87, 222 (1983).

14. W. M. Kriven, in Adv. in Ceramics, edited by A. H. Heuer and L. W. Hobbs (The American Ceramic Society, Columbus, Ohio 1984) pp. 64-77.

15. W. Mader and M. Rühle, Proc. 10th ICEM Meeting, 2, 101 (1982).

16. W. Mader and M. Rühle, Inst. Phys. Conf. Ser. No. 68 (EMAG 1983) 385 (1983).

17. C. E. Inglis, Trans. Instron Nov. Archit., 55, 219 (1913).

18. J. D. Eshelby, Proc. Roy. Soc. A, 241, 376 (1957), 241, 561 (1957).

19. S. Timoshenko and J. N. Goodier, Theory of Elasticity, McGraw Hill, New York (1961).

20. A. Kelly, Strong Solids, Clarendon Press, Oxford, U.K. (1966).

21. J. Schijve, "Analysis of the Fatigue Phenomenon in Aluminium Alloys, Technical Report M2122, N.A.A.R.I. Amsterdam (1964).

22. A. A. Griffith, Phil. Trans. R. Soc., A221, 163 (1920).

23. S. M. Wiederhorn, B. J. Hockey, and D. E. Roberts, Phil. Mag., 28, 783 (1973).

24. S. M. Wiederhorn, J. Am. Ceram. Soc., **52**, 485 (1969).

25. M. Iwasa, T. Veno, and R. C. Bradt, J. Soc. Mater. Sci., Japan 30, 1001 (1981).

26. H. L. Cox, Br. J. Appl. Phys., **3**, 72 (1952)

27. M. Rühle, A. Strecker, D. Waidlich, and B. Kraus, in Science and Technology of Zirconia II, edited by N. Claussen, M. Rühle, and A. H. Heuer (The American Ceramic Society, Columbus, Ohio 1984) pp. 256-274; L. H. Schoenlein, M. Rühle, and A. H. Heuer, ibid., pp. 275-282; W. M. Kriven, ibid., pp. 64-77.

28. R. G. Company, R. E. Smallman, and M. H. Loretto, Metal Science, 261 (1976).

29. R. G. Company, M. H. Loretto, and R. E. Smallman, Metal Science, 253 (1976).

30. P. Angelini and W. Mader, in Proceedings of the Am. Electron Microscopy Society of America, edited by G. W. Bailey, San Francisco Press, San Francisco, CA (1986) pp. 498-499.

31. C. H. Hsueh, A. G. Evans, R. M. Cannon, and R. J. Brook Acta Metall., **34**, 927 (1986).

32. B. R. Lawn, B. J. Hockey, and S. M. Wiederhorn, J. Mat. Sci., **15**, 1207 (1980).

33. B. J. Hockey and B. R. Lawn, J. Mat. Sci., **10**, 1275 (1975).

34. B. R. Lawn and M. V. Swain, J. Mat. Sci., **10**, 113 (1975).

ON PREVALENT WHISKER TOUGHENING MECHANISMS IN CERAMICS

A.G. Evans, M. Rühle, B.J. Dalgleish, M.D. Thouless
Materials Program, College of Engineering
University of California Santa Barbara, California   93106

ABSTRACT
Some aspects of whisker toughening are reviewed. It is shown that several important toughened materials have a toughness dominated by the nonlinear bridging of intact whiskers. Such toughening is demonstrated to depend sensitively on the relative fracture resistance properties of the whiskers, the interface and the matrix. It is also shown that, when the interface fracture resistance is low, the frictional sliding behavior of the previously debonded interface and the whisker strength distribution exert a major influence on toughness, in accordance with pull-out phenomena.

INTRODUCTION
The incorporation of whiskers into a brittle matrix can be envisioned to enhance the toughness in accordance with several mechanisms: crack deflection,[1] microcracking[2] and crack bridging.[3] Each of these mechanisms has very different microstructural requirements. It is imperative therefore for the development of materials with optimal toughness that the prevalent mechanisms operative in various composite systems be distinguished and quantified. Recent progress toward this objective has been substantiatal.[4,5,6] This progress is reviewed in the present article and prospects for the future are examined.

For the preliminary development of basic reinforcement concepts it is insightful to examine rule-of-mixtures (ROM) toughening as a reference and to document deviations from this rule. Two extreme possibilities are typically involved and occur when the reinforcements are either brittle or tough. Examples of the former are Al_2O_3 or Si_3N_4 toughened with SiC whiskers. These examples are remarkable because the reinforcement has a lower toughness than the matrix and yet the composite can have a toughness greater than three times that of the matrix, indicative of large positive deviations from ROM toughening (Fig. 1). Clearly, for such behavior to obtain nonlinear effects must be involved.[7] It will be demonstrated that these nonlinearities occur at the whisker/matrix interface, by virtue of 'weak' interphases that can be encouraged to form within this region. Less remarkable, but equally as important technologically, are systems that have

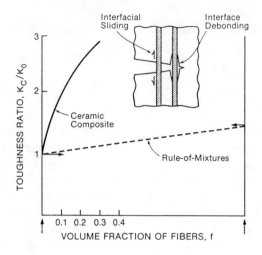

Fig. 1 A schematic comparison between a rule–of–mixture toughness
and positive deviations obtained in various whisker reinforced
systems.

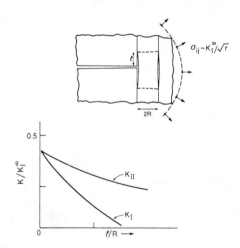

Fig. 2 A schematic of trends in mixed mode whisker debonding, occurring
at an angle $\theta = \pi/2$ with the crack plane.

small deviations from ROM toughening. These include glasses reinforced with SiC or Si_3N_4 whiskers and ceramics reinforced with metal whiskers. In these instances, strong interfaces are essential to the development of high toughness.[3,7]

The mechanical behavior of interfaces is still a poorly documented phenomenon, despite its critical importance to the properties of brittle matrix composites. Present understanding of fracture at interfaces is briefly addressed prior to consideration of the behavior of whisker reinforced systems.

INTERFACE FAILURE

Mechanics Considerations

In most situations pertinent to the failure of whisker reinforced ceramics, the interface is subject to mixed mode loading (Fig. 2). When a crack first contacts a whisker interface, competition between mode I growth across the whisker and mixed mode growth along the interface is established (Fig. 2). The nature of this competition is reasonably well-established when the elastic constants of the matrix and whiskers are essentially the same and the local response is elastic.[1] Then, in the whisker, at $\theta = 0$ (Fig. 2)

$$K_I/K_I^\infty = 1 \ : \ K_{II}/K_I^\infty = 0 \tag{1}$$

and along the interface, at $\theta = \pi/2$

$$K_I/K_I^\infty = 0.41$$

$$K_{II}/K_I^\infty = 0.40$$

such that

$$K \equiv \sqrt{K_I^2 + K_{II}^2} = 0.57 \ K_I^\infty, \tag{2}$$

where K_I^∞ is the stress intensity ascertained from the applied loads. Furthermore, should the crack extend along the interface in preference to propagation across the whisker, it is apparent that K must decrease with crack extension, while K_{II}/K_I increases. However, detailed results are not yet available. Such analyses of K, while insightful, have the limitation that crack growth preferences cannot be predicted without an adequate fracture criterion. In particular, mixed mode fracture envelopes need to be measured for typical material systems before predictions become feasible. Such measurements are in progress, using the test specimen depicted in Fig. 3, but data are still sparse.

In the absence of data, the only available recourse is the use of simplified debonding calculations, based on the energy release rate G (Fig. 4). Such calculations[8] suggest that the extent of debonding, ℓ, varies inversely with the interface fracture toughness, G_i (Fig. 4), in approximate accordance with; $\ell/R \sim (1/6) G_m/G_i$, where m refers to the matrix and R is the whisker radius. However, much additional study of this critical phenomenon is essential to further understanding of whisker toughening. In this context, it is noted that the basic mechanics of fracture have never been established when the elastic properties differ across the interface and when a thin interphase exists.

Observations

Experimental evidence of extreme interfacial responses have been obtained. Crack blunting has been observed at cracks extending from Al_2O_3 into Au, Nb[9] and Al, while debonding has been noted between Al_2O_3 and SiC (Fig. 5).[4] However, consistent observation of these extremes is elusive. In some cases, interfacial failure has been noted in Al_2O_3/Nb[9] and Al_2O_3/Au, while interfacial integrity has been observed in Al_2O_3/SiC.[4] The specific causes of such variable behavior have yet to be understood. Presently, only circumstantial evidence exists. This evidence suggests that direct bonding[10] generally provides 'strong' interfaces, while the presence of an amorphous interphase results in 'weak' interfaces.[11] Clearly, a concerted, broadly based study of interface strength and toughness is essential to further, basic understanding of the interface response to cracks.

TOUGHENING MECHANISMS

A preliminary characterization of whisker toughening mechanisms may be based on the 'strength' of the interface. When the interface is well-bonded and strong, toughening requires that one of two conditions obtain: either the whisker must be ductile and have a low modulus, causing local entrapment, blunting and bridging (Fig. 6), or it must have a sufficiently high modulus that it repels and deflects the crack (Fig. 7). When the interface is 'weak' and thus susceptible to debonding and sliding, toughening by crack bridging (Fig. 8) is possible and the extent of toughening is strongly influenced by the toughness properties of the interface and the whiskers. Residual stresses are also important, by virtue of

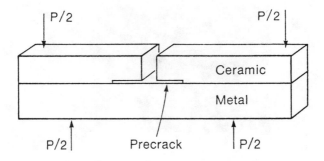

Fig. 3 A test procedure for evaluating mixed mode interfacial fracture loci.

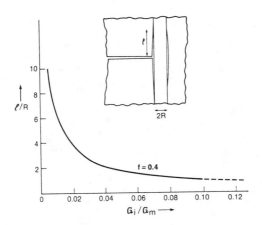

Fig. 4 Trends in relative debond length ℓ/R with relative debond resistance, G_i/G_m.

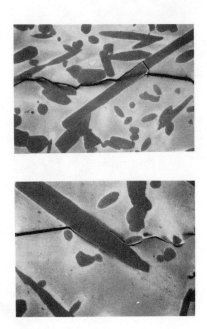

Fig. 5 Debonded whiskers within a bridging zone, observed in Al_2O_3/SiC.

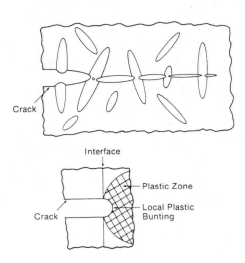

Fig. 6 Plastic deformation and blunting of metal whiskers intercepted by a matrix crack.

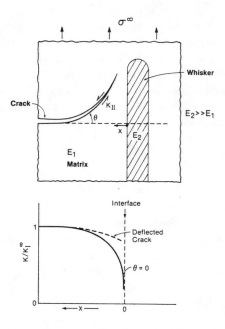

Fig. 7 Variations in stress intensity for a matrix crack in the vicinity
of a high modulus whisker.

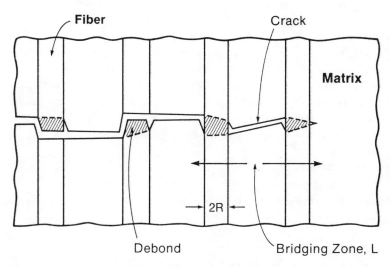

Fig. 8 A schematic of the coupled processes of interfacial debonding
and crack bridging.

their effect on interfacial debonding and whisker fracture, and may in some
cases have sufficient intensity to cause toughening by a microcrack process
zone mechanism.[7] A complete view of toughening would require that each of
these possibilities be analyzed and the predictions compared with experiment. Such a comprehensive review is not attempted. Instead, emphasis is
placed on the bridging mechanism because this mechanism seems to be
prevalent in several important whisker reinforced ceramics.[4]

Elastic Bridging

When limited debonding occurs, subsequent whisker failure initiates
relatively close to the matrix crack plane[4] (Fig. 9). Then, the whiskers
exert essentially linear elastic tractions on the crack surface, governed
by the elasticity of the system. For this case, explicit solutions for the
toughness have been derived[5,12,13] expressible in terms of either the
non-dimensional bridging zone size, L/R (Fig. 10a) or the strength
properties of the whiskers, S (Fig. 10b). Comparison between theory and
experiment is most effectively achieved in terms of the bridging zone size,
which can be measured using a variety of high resolution techniques[4] (Fig.
5). A comparison for Al_2O_3 materials reinforced with SiC indicates that
the elastic bridging model predicts the behavior of these composites with
reasonable precision (Fig. 10b). However, for purposes of toughness
optimization, the significance of the strength parameter S requires
clarification and definition. Available observations indicate that the
whiskers typically fail from the end of the debond zone[4] (Fig. 9). The
stress at whisker failure, S, is thus governed by the debond length ℓ and
the toughness of the whisker, K_w. This specific whisker failure process
has not been evaluated from either experimental or theoretical models and
consequently, explicit relationships between S, K_w and ℓ (or G_i/G_m) are not
yet available. Analogy with comparable axisymmetric fracture problems,
such as cone cracking,[14] would suggest the parametric form;

$$S \sim (K_w/\sqrt{R})(\ell/R)^{3/2} \tag{3}$$

indicative of strong effects of debond length ℓ and of whisker toughness,
K_w.

Nonlinear Bridging

When debonding occurs readily and whisker failure exhibits appreciable
spatial distribution about the matrix crack plane, as dictated by statis-

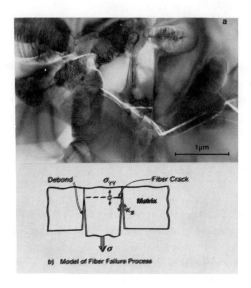

Fig. 9   Whisker failures close to the matrix crack plane.

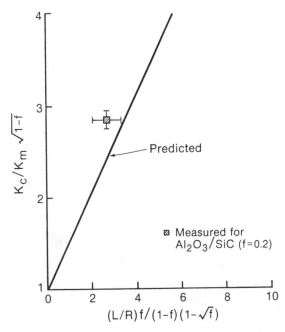

Fig. 10a   Trends in toughness associated with elastic bridging, bridge length, L.

tical variability in whisker strength, the tractions σ exerted by the
whiskers on the crack surface become nonlinear[15] (Fig. 11). The rising
portion of the traction function is dictated largely by the still intact
whiskers in the immediate crack wake, while the declining portion is
dominated by the failed whiskers as they 'pull-out' against the frictional
resistance of the debonded interface. The change in toughness provided by
the whiskers ΔG_c is asymptotically governed by the area encompassed within
the traction curve;[3]

$$\Delta G_c = 2f \int_0^{u_*} \sigma(u)du \qquad (4)$$

where u* is the crack opening at either the crack mouth or the end of the
bridging zone (whichever is the larger). It is thus apparent (Fig. 12)
that nonlinear effects (which become possible when the debond lengths are
large) can result in substantial composite toughness. Optimization of such
nonlinear effects emerges as a vital aspect of the development of high
toughness whisker reinforced ceramics.

Preliminary analysis has been conducted[15] for the case wherein the
interfacial debond resistance is very small, resulting in large debond
lengths. In this case, the debonded interface can be characterized by an
essentially unique shear resistance, τ, that depends on the whisker
topology, residual stress, and the friction coefficient.[8] Then, statistical analysis of the distribution of most probable whisker fracture
sites[15] allows prediction of the traction function (Fig. 11). Computation
of the asymptotic toughening suggests the trend

$$\Delta G_c \sim S_o^{3m/(m+1)} (R/\tau)^{1-3/(m+1)} f \qquad (5)$$

where S_o is the scale parameter in the statistical distribution of fiber
strengths and m is the shape parameter that governs the strength variability.[16] Hence, high strength whiskers with wide statistical variability, having large radii and a low interfacial shear resistance provide
the optimum asymptotic toughness.

Another feature of fracture behavior in materials having weakly bonded
interfaces is the inevitable existence of continuously rising resistance
curves. Such behavior emerges because the pull-out lengths for short
cracks exceed the crack mouth opening. However, specific trends remain to
be evaluated. Furthermore, an important behaviorial regime that has yet to
be evaluated occurs when crack tip debonding allows strong whiskers to

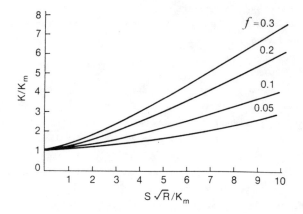

Fig. 10b Trends in toughness associated with elastic bridging, whisker strength, S.

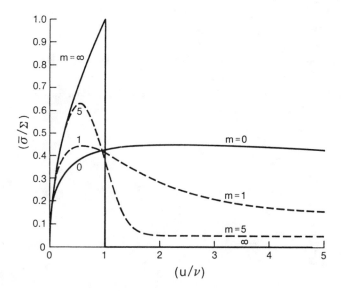

Fig. 11 Nonlinear bridging effects caused by pull-out after whisker failure.

fracture within the debond region, while others fail at the end of the debond. This regime is expected to have considerable practical utility and thus merits future emphasis.

CONCLUDING REMARKS

A brief summary of the current status of understanding regarding the whisker toughening of ceramics has been presented. The emphasis has been on the bridging mechanism, because several improtant whiskers toughened materials appear to be toughened in accordance with this mechanism. However, the prevalence of other mechanisms in certain whisker toughened materials (e.g., crack deflection in glass matrix systems) is certainly not precluded by the present emphasis.

When bridging processes dominate, interface debonding appears to be a prerequisite for appreciable toughening, as needed to prevent whisker failure at the crack tip. In the examples cited, debonding involves the presence of a thin amorphous interphase. It remains, however, to identify interphase characteristics that optimize the debonding process.

Weakly bonded interfaces that result in extensive tip debonding provide additional 'pull-out' contributions to toughness. The enhancement in toughness is governed by the frictional shear resistance of the debonded interface and by the whisker strength distribution. Typically, small values of the shear resistance, large mean strengths and large statistical deviations in whisker strength are needed to cause appreciably enhanced crack growth resistance. Approaches for generating such interfaces, that are also stable at elevated temperatures, provide a major challenge for the development of tough, whisker reinforced ceramics.

REFERENCES

1. K.T. Faber and A.G. Evans, Acta Met., <u>31</u> (1983), 565.

2. A.G. Evans and K.T. Faber, J. Amer. Ceram. Soc., <u>67</u> (1984), 255.

3. A.G. Evans and R.M. McMeeking, Acta Met., <u>34</u> (1986), 2435.

4. M. Rühle, B.J. Dalgleish, and A.G. Evans, Scripta Met., in press.

5. B. Budiansky, Micromechanics II, Tenth National Congress of Applied Mechanics, Austin, Texas, 1986.

6. D.B. Marshall and B.N. Cox, Acta Met., to be published.

7. A.G. Evans, "Ceramic Microstructure: The Role of Interfaces," (ed. J.A. Pask and A.G. Evans), Plenum, NY, (1986), in press.

8. B. Budiansky, J.W. Hutchinson and A.G. Evans, J. Mech. Phys. Solids, <u>2</u> (1986), 167.

9. A.G. Evans, M.C. Lu, M. Rühle and S. Schmauder, Acta Met., <u>34</u> (1986), 1643.

10. M. Rühle and W. Mader, J. Mater. Sci., in press.

11. I. Aksay, "Ceramic Microstructure: The Role of Interfaces," (ed. J.A. Park and A.G. Evans), Plenum, NY (1986) in press.

12. B. Budiansky, J. Amazigo and A.G. Evans, to be published.

13. F.R.G. Rose, J. Mech. Phys. Solids, in press.

14. B.R. Lawn and T.R. Wilshaw, Fracture of Brittle Solids, Cambridge Univ Press (1975).

15. M.D Thouless and A.G. Evans, Acta Met., to be published.

16. R.L. Stewart, K. Chyung, M.P. Taylor and R.F. Cooper, Fracture Mechanics of Ceramics (ed. R.C. Bradt, F.F. Lange, A.G. Evans and D.P.H. Hasselman), Plenum, NY (1986), Vol. 7, p. 33.

MICROSTRUCTURE DEVELOPMENT IN Si_3N_4-BASED COMPOSITES

S.T. BULJAN AND G. ZILBERSTEIN
GTE Laboratories Incorporated, 40 Sylvan Road, Waltham, MA 02254

ABSTRACT

The success and predictability of mechanical property improvements in ceramic matrix composites strongly depends on the ability to generate specific spacial arrangements and maintain the properties of the constituents. Dispersoids added to improve the toughness and strength of the ceramic base can mechanically or chemically alter the development of the matrix microstructure during consolidation and cause the properties to deviate from expected values. The present study examines and compares the development of a Si_3N_4-based composite microstructure with the chemically "active" TiC and "inert" SiC dispersoids; it further examines the effect of the resulting microstructures on the composite fracture toughness.

INTRODUCTION

Modifying ceramics through addition of dispersoids offers a potential for considerable improvements in both fracture toughness and strength. Recent literature indicates that crack deflection which produces twist and tilt of an advancing crack front is an effective toughening mechanism. Reports also predict increases in fracture toughness of the composites with increasing content of dispersoid phase regardless of its grain size or sign of thermal expansion mismatch.[1-3] The degree of toughening is strongly affected by the shape of the dispersoid, whiskers being several times more effective toughening agents than equiaxed particles.[2] Although results obtained for some glass matrix systems seem to bear out the projected trends, typical results for ceramic matrix composites more often than not deviate from such projections. These differences stem from the fact that fracture mechanics calculations treat matrix material as a homogeneous continuum. The microstructure of polycrystalline and often polyphase ceramic matrices, however, require special consideration in composite design.

Ceramics rarely fracture in an exclusively transgranular mode. The cracks in ceramics fracturing partially and substantially intergranularly are deflected by the grains, and it is reasonable to expect that fracture toughness of such materials is controlled by the grain size. The development of microstructure is influenced by processing and physicochemical interactions of constituents during sintering. Interactions with dispersoids may chemically or mechanically influence the development of matrix microstructure and increase the uncertainty in attainment of properties predicted by simplified models. While it is apparent that extensive chemical reactions will substantially alter the phase assembly and properties, it may also be possible for limited reactions which contribute to minor change in composition of the glass phase to alter crystallization kinetics and also affect properties. The present study examines and compares the development of Si_3N_4-based composite microstructure with the chemically "active" TiC and "inert" SiC dispersoids and its effect on fracture toughness.

PRECURSOR MATERIALS AND PROCESSING

Table I gives the results of chemical analysis of impurities found in major precursor powders used in this study, Si_3N_4 (SN502 GTE Prod. Corp., Towanda, PA), TiC (Cerac, Inc., Milwaukee, Wi) and SiC-Silar™ SC-9, (Arco Corp.).

Table I: Impurities in Precursor Powders

Precursor Powder	Impurities, Wt. %											
	Metals									Nonmetals		
	Al	Mg	Ca	Fe	Cr	Ni	Mn	Mo	Free Si	Free C	O_2	Cl
	Emission Spectroscopy								XRD	CTC, Leco	NAA	XRF
Si_3N_4	<0.01	ND	—	<0.05	ND	ND	ND	<0.005	<1.0	ND	2.0	<0.05
SiC	<0.1	0.1–0.3	0.1–0.3	<0.1	<0.1	<0.1	0.1–0.3	ND	ND	0.36	1.78	ND
TiC	0.05	ND	0.01	0.05	0.01	ND	0.001	ND	ND	0.1	0.27	ND

The Si_3N_4 powder typically contained $\approx$ 94% αSi_3N_4 and $\approx$5% β-Si_3N_4, with an initial surface area of 4-5 m^2/g. The particles had acicular morphology. After milling and blending, the Si_3N_4 surface area usually increased to 7.0-8.0 m^2/g, which corresponded to $\approx$0.25 μm, average Si_3N_4 grain size. The TiC powder consisted of equiaxed particles of $\approx$1.5 μm average diameter. The SiC used in this study was whisker material with an initial average length $\overline{l}$ = 17.5 ($\pm$12.06) μm, average cross sectional "diameter" $\overline{d}$=0.53($\pm$0.19) μm, and the aspect ratio l/d$\approx$33.0. During processing, the average SiC aspect ratio was reduced to an ultimate value of $\approx$6.0 in the densified composite.

All materials were densified using 6 w/o Y_2O_3-2 w/o Al_2O_3 sintering aids, by hot pressing in argon atmosphere at 1725-1750°C to full density. The dispersoid content varied from zero to 30 v/o, in 10 v/o increments.

Si_3N_4-TiC COMPOSITE

The microstructure of Si_3N_4-30 v/o TiC particulate composite prepared by a standard processing schedule is shown in Figure 1a. Holding at the densification temperature for an extended time reveals the presence of a reaction product zone around every TiC particle (Figure 1b). It has been shown[4,5] that this reaction zone is composed of SiC and $TiC_{0.5}N_{0.5}$ crystals which form at elevated temperatures, according to the reaction:

$$Si_3N_4 \pm 6\,TiC = 3\,SiC + 6\,TiC_{0.5}N_{0.5} + ½\,N_2 \tag{1}$$

At the densification temperature, the reaction can ($\Delta G_T^\circ \approx$ -30,000 cal/mole) proceed until all TiC is consumed. Hence, if the reaction (1) is carried to completion the properties expected from such a composite, which were based on the properties of the starting materials and their respective (volume) quantities in the composite formulation, will not be realized because of substantial changes in the phase composition. Appropriate adjustment of processing parameters (standard processing schedule, Fig. 1a) produces a composite with limited reaction which essentially preserves the dispersoid properties and confines the reaction to a very narrow zone at the matrix-dispersoid interphase boundary. This limited reaction, however, affects the matrix microstructure development. Figure 2 shows the TEM photomicrograph of the dispersoid-free (a) and the 30 v/o TiC-containing Si_3N_4 composite (b) having the same amount of sintering aids and hot pressed identically. Figure 2 shows that the Si_3N_4 microstructure of the TiC-free material is characterized by larger Si_3N_4 grain sizes and a substantially broader Si_3N_4 grain size distribution than that in the TiC-containing composite. The Si_3N_4 grain size and distribution were evaluated using quantitative microscopy and are given in Figure 3 as relationships between the Si_3N_4 equivalent grain diameter* and its volume percent of the total Si_3N_4 content in each material.

---

* diameter of a circle whose area is equal to the grain area measured from the photomicrograph

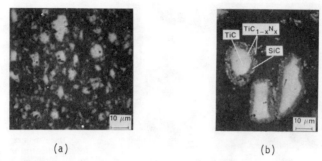

(a)　　　　　　　　　　　　(b)

Figure 1:　Microstructure of Si₃N₄-TiC composites prepared by a standard (a) and an extended (b) time processing

(a)　　　　　　　　　　　　(b)

Figure 2:　TEM bright field images of the dispersoid-free (a) and 30 v/o TiC-containing Si₃N₄ composites (b)

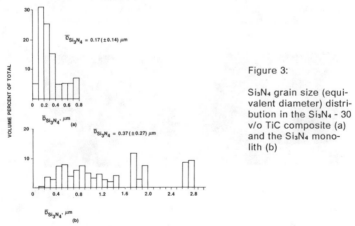

Figure 3:

Si₃N₄ grain size (equivalent diameter) distribution in the Si₃N₄ - 30 v/o TiC composite (a) and the Si₃N₄ monolith (b)

The observed difference in Si₃N₄ grain sizes and distributions could be attributed to the differences in the Si₃N₄ solution precipitation and growth behavior which are apparently influenced by Ti present in the glass of the Si₃N₄-TiC composite as witnessed by the EDX spectrum of the glass region obtained in the STEM (Figure 4).

276

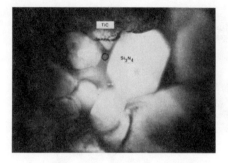

Figure 4:

Energy dispersive
x-ray spectrum of
the glass region
(indicated by a
circle on the upper
photomicrograph) in
the Si_3N_4-30 v/o
TiC composite.

One of the goals pursued in the design of Si_3N_4-TiC composites was to improve the material's fracture toughness by one of the "crack-TiC particle" interaction mechanisms. Although substantial improvements in hardness and wear resistance were achieved, the composite's fracture toughness remained at the level of the Si_3N_4-monolith at best. Crack propagation patterns in the Si_3N_4 monolith and the Si_3N_4 - 30 v/o TiC hot pressed composites were examined on the polished specimen surfaces precracked with a single Knoop indenter using 100 N load (Figure 5). The patterns show that amplitude of crack deflection in the composite matrix is considerably smaller than that observed in the monolith.

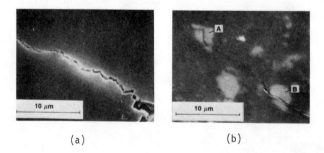

(a)   (b)

Figure 5: Crack propagation patterns in Si_3N_4 (a) and Si_3N_4-TiC composite (b) (after [6])

The matrix of the Si_3N_4-TiC composite is characterized by the Si_3N_4 grain size that is less than half that of the monolith, and the crack propagation path through the fine grain material, fracturing intergranularly, should be less tortuous than in a coarser grain counterpart. Figure 6 gives an example of an intergranular crack path in the Si_3N_4 - 30 v/o TiC composite.

Hence the fracture toughness of the Si₃N₄ matrix appears to be affected by the grain size. The change in grain aspect ratio would intensify the effects of crack deflection considerably. According to the proposed relationship[7] in a material which fractures in a typically intergranular mode the expected change in fracture toughness (ΔK_{IC}) due to grain size is

$$\Delta K_{IC} = CK_{IC}^{\circ} (D/D^{\circ}-1) \tag{2}$$

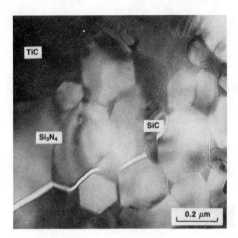

Figure 6: Crack propagation pattern in the Si₃N₄ 30 v/o TiC composite

where C is a proportionality factor, and K_{IC}° represents in this case the fracture toughness of the Si₃N₄ monolith with the D° average grain size. A typical indentation fracture toughness value measured on hot pressed Si₃N₄ monolith is K_{IC}° = 4.5 MPa•m$^{1/2}$; the average Si₃N₄ grain size D° = 0.37 μm. Then the expected change in K_{IC} due to the grain size change, ΔK_{IC} at $D_{Si_3N_4}$ = 0.17 μm for such Si₃N₄ monolith is -0.6 MPa•m$^{1/2}$, (C = 0.25) a reduction in fracture toughness. Assuming that there is no major change in the relative amount of energy required to fracture the intergranular phase or in the amount of an intergranular fracture, the fracture toughness of 4.5 MPa•m$^{1/2}$ measured for Si₃N₄ - 30 v/o TiC composite is a product of fracture toughness provided by the crack deflection in Si₃N₄ matrix and by TiC dispersoid, where TiC provides 40% toughness increase over the Si₃N₄ monolith of finer (0.17 μm) grain size. Although it is clear from the discussion presented above that incorporation of a hard dispersoid of a given (1.5 μm) grain size into Si₃N₄ matrix could potentially provide an increase in the composite fracture toughness, this effect is cancelled by the lower fracture toughness of refined Si₃N₄ grain size matrix due to the Si₃N₄ + TiC reaction during densification.

Si₃N₄-SiC WHISKER COMPOSITES

Si₃N₄ matrices reinforced with whisker dispersoids offer greater opportunities in design of tough and wear-resistant ceramic materials. It has been proposed[2] that incorporation of rod-like ceramic particles into a continuous matrix (glass) may result in a four-fold increase in a composite's fracture toughness as compared to sphere-shaped dispersoids which, at the limit, offer only two-fold increase in toughness. Although this model did not consider a matrix material as a polycrystalline body with glass at the boundaries, it showed a high potential for ceramic matrix toughening by whisker additions. The presence of whiskers retards densification kinetics and requires adjustment of the sintering schedule to obtain fully dense materials. Such adjustments (longer sintering time, higher temperature) are expected to affect matrix microstructure development. A major consideration in favor of using SiC whiskers to toughen Si_3N_4 matrix was the SiC chemical inertness relative to the Si_3N_4 during densification under carefully selected processing conditions. While serving as a toughening component, the SiC whiskers will not interfere chemically with Si_3N_4 grain growth, thus allowing attainment of an optimized matrix microstructure and related properties through the high temperature processing. In a material with typical intergranular fracture, the energy expended to propagate a crack appears to be directly proportional to the amplitude and frequency of crack deflection by the grains. The amplitude and frequency are defined by the grain dimensions. Hence if the material lends itself to controlled grain growth by processing, its fracture toughness could be controlled as well. Taking an average grain size $D°$ and associated fracture toughness $K_{IC}°$ obtained in a material by using standard processing as a reference, the expected change in K_{IC}, the ΔK_{IC}, should be a function of a grain size obtained by modified processing and can be expressed as shown in Eq. (2).

To test the proposed behavior, two sets of Si_3N_4-(0-30) v/o SiC whisker composites were hot pressed at the same temperature using two heating schedules. The first schedule was based on the shortest time necessary to achieve a complete α-β Si_3N_4 conversion and 100% theoretical density. The second one was to extend time to 400 min. for each composition to provide the equivalent conditions for Si_3N_4 grain growth. The resulting microstructures were examined by quantitative metallographic analysis, and Si_3N_4 grain size and distribution determined from STEM micrographs at 25,000X magnification. Figure 7 gives an example of Si_3N_4-20 v/o SiC whisker microstructure and Si_3N_4 grain size distribution for composites hot pressed for 210 min. (a) and 400 min. (b). Table II contains the quantitative metallographic analysis data on all examined compositions and the associated fracture toughness values.

Table II: Microstructure and Fracture Toughness Data of Si_3N_4-(0-30) v/o SiCw Composites Hot Pressed by Standard and Extended Time Schedules

N	Composition v/o	HP Time min	$\overline{D}$. Si_3N_4 µm	$\pm \sigma_n^D$	K_{IC} MPa•$m^{1/2}$
1	AY6	90	0.37	0.27	4.70 ± 0.30
2	AY6	400	0.59	0.41	5.40 ± 0.50
3	AY6-10 v/o SiCw	110	0.36	0.24	4.40 ± 0.10
4	AY6-10 v/o SiCw	400	0.47	0.30	6.00 ± 0.10
5	AY6-20 v/o SiCw	210	0.24	0.14	4.80 ± 0.30
6	AY6-20 v/o SiCw	400	0.49	0.32	5.90 ± 0.80
7	AY6-30 v/o SiCw	400	0.36	0.24	6.40 ± 0.50

Figure 7 and Table II show that an increased fracture toughness of composites hot pressed for 400 min. is clearly due to the combined effect of a well developed matrix microstructure and the additional toughening effect of SiC whiskers. In order to assess the applicability of the proposed relationship between the grain size and the resulting composite fracture toughness to predict an expected composite fracture toughness using the microstructural parameters, the K_{IC} of the composites listed in Table II were calculated. Figure 8 compares the actual measured and the calculated K_{IC} data vs v/o SiC for all seven compositions. Figure 8 indicates that the dependence resulting from using the calculated K_{IC} values (the dashed lines) approximates to a substantial degree the one based on actual measurements, despite the assumption that both matrix and whisker dispersoid grains are equiaxed. Although the average SiC whisker aspect ratio is l/d = 6.0 in the hot pressed composites, the average equivalent SiC grain diameter $D_{sic} = 1.95$ µm was used in the calculations. The actual shape distribution of Si_3N_4 grains in AY6-20 v/o SiC hot pressed for 210' and 400' was evaluated using quantitative microscopy and is shown in Figure 9. The form factor varies between zero and 1.0, with F = 1.0 corresponding to an equiaxed or circular grain shape and 0.0 < F < 1.0 describing a deviation from a circle toward more elongated shapes. The average aspect ratio for Si₃N₄ grains calculated using the data from Figures 9 and 7 was l/d = 1.5 in the 210 min. specimen and l/d = 1.8 in the 400 min. specimen. Since these data are based on a large number of measurements, they apparently reflect actual Si₃N₄ grain growth kinetics during liquid phase sintering. These data indicate that the Si₃N₄ grain growth rate is rather slow, that the grain growth is, for the most part, planar, and that the Si₃N₄ grain aspect ratio is maintained at a low and constant level, independent of sintering time.

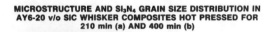

MICROSTRUCTURE AND Si₃N₄ GRAIN SIZE DISTRIBUTION IN AY6-20 v/o SIC WHISKER COMPOSITES HOT PRESSED FOR 210 min (a) AND 400 min (b)

Figure 7:

TEM bright field images and the Si₃N₄ grain size (equivalent diameter) distributions in the Si₃N₄-20 v/o SiC whisker composites hot pressed for 210 min. (a) and 400 min. (b)

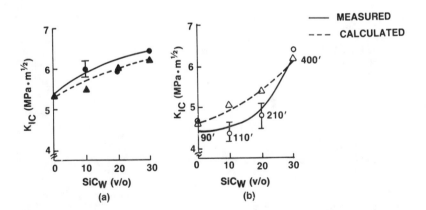

Figure 8: Comparison between the measured and the calculated K_{IC} values vs SiCw content for AY6-(0-30) v/o SiC whisker composites (a) 400 min. HP; (b) at minimum time to full density.

280

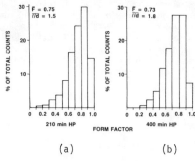

(a)   (b)

Figure 9:

Shape distribution of Si_3N_4 grains in AY6-20 v/o SiCw composites hot pressed for 210 min. (a) and 400 min. (b)

Based on these observations, it is apparent that in the case of discontinuous composites that fracture in an intergranular mode the matrix grain size influences the choice of the dispersoid size and therefore the relationship between these two should be taken into consideration.

CONCLUSIONS

Si_3N_4 microstructural development during densification in the presence of TiC and SiC whisker dispersoids was examined. It was shown that the presence of a TiC dispersoid which is chemically reactive with a Si_3N_4 matrix leads to substantial matrix grain refinement as compared to the Si_3N_4 monolith. Fine- grain Si_3N_4 matrix microstructure was shown to be responsible for the lack of fracture toughness improvements in Si_3N_4-TiC composites as compared to the Si_3N_4 monolith. Chemical inertness of SiC relative to Si_3N_4 allows controlled microstructure development in Si_3N_4-SiC whisker composites. Fracture toughness of SiC whisker-reinforced Si_3N_4 matrix composites was shown to be a combined result of a well-developed matrix microstructure and the additional toughening effect of SiC whiskers.

ACKNOWLEDGMENT

Research sponsored in part by U.S. Dept. of Energy Ceramic Technology for Advanced Heat Engines Project Monitored through Oak Ridge National Laboratory; Contract No. DE-AC05-840R-21400, Martin Marietta. Authors also acknowledge experimental assistance of M. Katsoulakos, K. Ostreicher and T. Emma.

REFERENCES

1.   F.F. Lange, "The Interaction of a Crack Front with a Second Phase Dispersion," Phil. Mag. 22: 983-992 (1980).

2.   K.T. Faber and A.G. Evans, "Crack Deflection Process - I Theory," Acta. Metall., 31 (4), 565 (1983).

3.   J.C. Swearengen, et al., "Fracture Toughness of Reinforced Glasses," 973-987, Fracture Mechanics of Ceramics, Vol. 4 (R.C. Bradt, et al., ed.) Plenum Press, New York (1978).

4.   G. Zilberstein and S.T. Buljan, "Characterization of Matrix-Dispersoid Reactions in Si_3N_4-TiC Composites," Advances in Materials Characterization II, Materials Science Research, Vol. 19, 389-402, (R.L. Snyder, et al., ed.) Plenum Press, New York (1985).

5. S.T. Buljan and G. Zilberstein, "Effect of Impurities on Microstructure and Mechanical Properties of Si_3N_4-TiC Composites," "Tailoring of Multiphase and Composite Ceramics," Materials Science Research, vol. 20, 305-316 (R.E. Tressler, et al., ed.) Plenum Press, New York (1986).

6. J.G. Baldoni, et al., "Mechanical Properties and Wear Resistance of Silicon Nitride-Titanium Carbide Composites," "Tailoring of Multiphase and Composite Ceramics," Materials Science Research, vol. 20, 329-346 (R.E. Tressler, et al., ed.) Plenum Press, New York (1986).

7. S.T. Buljan, et al., "Si_3N_4-SiC Composites," Am. Cer. Soc. Bull., 66(2), 347-52 (1987).

DEFLECTION-TOUGHENED CORUNDUM-RUTILE COMPOSITES

SABURO HORI*, HISATSUGU KAJI*, MASAHIRO YOSHIMURA** AND SHIGEYUKI SŌMIYA**
* Kuraha Chemical Industry Co., Ltd., 16 Ochiai, Nishikimachi, Iwaki-Shi, Fukushima 974, Japan
** Research Laboratory of Engineering Materials, Tokyo Institute of Technology, 4259 Nagatsuta, Midori-Ku, Yokohama 227, Japan

ABSTRACT

Dense corundum-rutile composites containing plate-shaped corundum particles were prepared from the CVD Al_2O_3-TiO_2 powders by adding a small amount of sodium and sintering below ~1280°C. The fracture toughness increased up to 6.5 MPa·√m and the increment of toughness was proportional to the volume % of plate-shaped corundum particles. The importance of tilt deflection and crack bowing rather than the twist deflection was suggested in this system.

INTRODUCTION

Ceramics can be toughened by several mechanisms, among which most successful in the past decade were the transformation toughening and the microcrack toughening utilizing the phase change of ZrO_2 from tetragonal to monoclinic [1,2]. As these toughening phenomena are related to the transformation of ZrO_2 and therfore temperature-dependent, the fracture toughness usually becomes low at high temperatures, typically ≥700°C. Recently the toughening by crack deflection [3] is attracting much attention especially because it is insensitive to temperature. The crack deflection toughening can be achieved most effectively by incorporating rod- or disc-shaped particles of high aspect ratio into ceramic matrices. But it is very difficult to process these particles of high aspect ratio. It is desirable to grow these rod- or disc-shaped particles during the sintering from spherical starting powders.

In the Al_2O_3-TiO_2 system, aluminum titanate (Al_2TiO_5) is a stable compound only at temperatures higher than ~1300°C and the coexistence of corundum and rutile is expected as an equilibrium state at lower tempratures [4]. Okamura et al. [5] fabricated a two-phase structure consisting of corundum and rutile from TiO_2-coated Al_2O_3 powders, but the density achieved was only ~90% even with prolonged heating (1280°C, 20 h).

The present authors prepared codeposited Al_2O_3-TiO_2 powders consisting of γ or δ-Al_2O_3 and rutile TiO_2 by a flame CVD technique [6], which exhibited an excellent sinterability, to fabricate Al_2TiO_5 [7,8]. Even from these CVD Al_2O_3-TiO_2 powders, it was difficult to obtain fully dense ceramic materials consisting of corundum Al_2O_3 and rutile TiO_2. But by adding a small amount of sodium (Na) to the CVD powders, the sinterability was significantly improved and, rather unexpectedly, the growth of thin hexagonal plate-shaped corundum particles was observed.

This paper reports the growth of plate-shaped corundum particles in the corundum-rutile composites and the increase of fracture toughness associated with the plate-shaped particles.

EXPERIMENTAL PROCEDURES

Addition of small amount (0.025-0.12 wt%) of Na to the Al_2O_3-TiO_2 powders was achieved by two methods, one of which was the rather accidental contamination from a Na-containinng porous refractory material used as the inner wall of the reactor during the CVD reaction to produce the powders, and the other was addition of Na_2CO_3 after the CVD powder production. The Na-containing powders were isostatically pressed at 294 MPa into pellets (approximate sintered dimensions 10 mm^ϕ x 5 mm^t) and sintered in air at 1250°C for 1 to 14 h. Phases detected were corundum and rutile. After polishing, the pellets were subjected to bulk density measurement, fracture toughness (K_{IC}) measurement and microstructural observation. The K_{IC} was determined by the indentation fracture (IF) technique initially proposed by Evans and Charles and modified by Niihara [9] with Vickers load of 294 N. The microstructures of polished specimens were observed by scanning electron microscopy (SEM) using backscattered electron reverse image mode so that corundum and rutile could be easily distinguished. From the SEM pictures, volume fraction and average dimensions of plate-shaped particles were determined. Fullman's relations [10] were used to determine the average particle diameter and thickness.

RESULTS AND DISCUSION

Increase of Fracture Toughness by Microstructural Development

The Al_2O_3-TiO_2 powder (Al_2O_3/TiO_2 weight ratio 52.5/47.5) with 0.12 wt% Na added during the CVD reaction was sintered for 1 to 14 h at 1250°C. The densities achieved were ~97 % of theoretical with specimens sintered for 1 h and 99.0 - 99.7 % with those sintered for 3 to 14 h. As shown in Fig. 1, corundum particles of rectangular-shaped cross section were grown by extending the sintering (ageing) time, and the propagating cracks were deflected more as the corundum particles increased in size. By observing the fractured surfaces, the shape of these particles were confirmed to be a thin hexagonal plate. It should be noted that the development of plate-shaped particles was less significant inside the specimens than near the sintered surface.

The volume % and the average diameter and thickness (approximated as discs) of these plate-shaped corundum particles were shown in Fig. 2 as a function of ageing time. By extending the ageing time from 1 to 14 h, the vol% increased from 5 to 30 %, and the average diameter(ϕ) and thickness(t) increased from 8 μm^ϕ x 0.6 μm^t to 13.5 μm^ϕ x 1.6 μm^t. The fracture toughness (K_{IC}) also increased from 3 to 6.5 MPa·$\sqrt{m}$ by the development of these plate-shaped particles as shown in Fig. 3. The increment of toughness appeared to be proportional to the vol% of plate-shaped particles.

Faber and Evans [3] argued that the K_{IC} increase by crack deflection was mostly achieved by twist deflection and that the K_{IC} reaches an asymptotic level with ~10 vol% of dispersed particles. The theory does not agree very well with the experimental results. In this specific system, the tilt deflection and crack bowing may be the dominant mechanisms and the contribution of twist is probably less than in the theory.

Effects of Additives on Developmment of Plate-Shaped Particles

The Al_2O_3-TiO_2 powder (Al_2O_3/TiO_2 weight ratio 46.3/53.7) containing only 0.006 wt% Na was used as the initial powder for this study. Three kinds of powders were prepared; (1) without additives, (2) with Na_2CO_3 addition, and (3) with Na_2CO_3 and submicron α-Al_2O_3 (AMS-12, Sumitomo Aluminium Smelting Co.) additions. The submicron α-Al_2O_3 was added in the amount of 1 wt%. The total Na after the treatments was 0.025 wt% in case (2) and 0.035 wt% in case (3).

Table 1 summarizes the properties of the specimens sintered at 1250°C for 6 h from these powders. The microstructural observation revealed that the plate-shaped corundum particles grew only in the case (2), as shown in

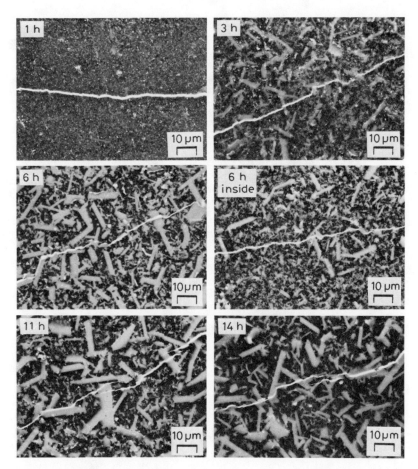

Fig. 1. Growth of plate-shaped corundum particles (appearing white) by
extention of ageing time at 1250°C and interactions between particles and
crack introduced by Vickers indentation in a corundum-rutile composite
(Al_2O_3/TiO_2 weight ratio 52.5/47.5) containing 0.12 wt% Na. The growth of
plate-shaped particles was less significant inside the specimens. The SEM
micrographs were taken by backscattered electron reverse image mode on the
polished specimens.

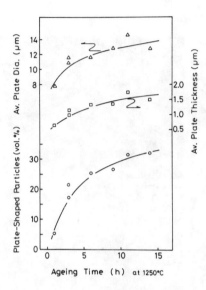

Fig. 2. Increase of vol% and dimensions of plate-shaped corundum particles as function of ageing time at 1250°C in a corundum-rutile composite (Al_2O_3/TiO_2 weight ratio 52.5/47.5) containing 0.12 wt% Na.

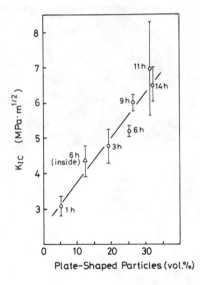

Fig. 3. Dependence of fracture toughness (K_{IC}) on vol% of plate-shaped corundum particles in a corundum-rutile composite (Al_2O_3/TiO_2 weight ratio 52.5/47.5) containing 0.12 wt% Na. The increment of K_{IC} appeared to be proportional to the vol%.

Table 1. Effects of Additives on Density and Fracture Toughness of
Corundum-Rutile Composites*

Case	Additive	Total Na (wt%)	Density (% TD)	K_{IC} (MPa·$\sqrt{m}$)
1	none	0.006	90.9	2.84 ± 0.09
2	Na_2CO_3	0.025	95.8	4.85 ± 0.40
3	Na_2CO_3 α-Al_2O_3 (1 wt%)	0.035	92.3	3.16 ± 0.50

* Al_2O_3/TiO_2 weight ratio 46.3/56.7, sintered at 1250°C for 6 h.

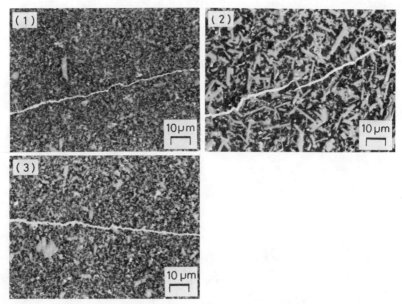

Fig. 4. Effects of additives on microstructure of corundum-rutile
composites (Al_2O_3-TiO_2 weight ratio 46.3/56.7). (1) without Na addition (Na
present as an impurity 0.006 wt%), (2) with Na_2CO_3 addition (0.025 wt% Na),
and (3) with Na_2CO_3 (0.035 wt% Na) and submicron α-Al_2O_3 (1 wt%).

Fig. 4. From these results, the liquid phase formation and the nucleation of plate-shaped particles are considered to be very important for the microstructural development.

A small amount (≈ 0.1 wt%) of Na is sufficient to form a liquid phase probably in the grain boundary between corundum and rutile, which promotes an Ostwald ripening (dissolution and precipitation) of corundum particles. In addition to the liquid phase, the nucleation of the plate-shaped corundum particles is important. As the addition of α-Al_2O_3 appeared to prevent the nucleation, the nucleation of plate-shaped corundum particles are strongly related to the transformation of γ or δ-Al_2O_3 to α-Al_2O_3. Also the metastable solid solution of Al_2O_3 in TiO_2 and of TiO_2 in Al_2O_3 in the CVD powder probably plays an important role to form the nulei of plate-shaped particles.

SAMMARY

Plate-shaped corundum paticles could be grown in corundum-rutile composites by an Ostwald ripening. Not only the Na-containing liquid phase, but also the nucleation associated with the phase change of $\gamma(\delta)$- to α-Al_2O_3 are important for the development of plate-shaped particles.

The fracture toughness increased significantly, the increment of which appeared to be proportional to the vol% of plate-shaped particles. It is suggested that the contirbution of twist deflection is less significant than those of tilt deflection and crack bowing.

Further study on the microstructureal development and the relationship between microstructure and mechanical properties will clarify the toughening phenomena in these corundum-rutile composites.

REFERENCES

1. R.C. Garvie, R.H. Hannink and R.T. Pascoe, Nature (London) 258, 703 (1975).
2. A.G. Evans, in Advances in Ceramics 12, edited by N. Claussen, M. Rühle and A.H. Heuer (The American Ceramic Society, Columbus, OH, 1984) pp. 193-212.
3. K.T. Faber and A.G. Evans, Acta Metall. 31, 565 (1983).
4. K. Hamano, Taikabutsu (Refractories) 27, 520 (1975).
5. H. Okamura, E.A. Barringer and H.K. Bowen, J. Am. Ceram. Soc. 69, C22 (1986).
6. S. Hori, Y. Ishii, M. Yoshimura and S. Sōmiya, Yogyo-Kyokai-Shi 94, 400 (1986).
7. S. Hori, R. Kurita, M. Yoshimura and S. Sōmiya, Int. J. High Tech. Ceram. 1, 59 (1985).
8. S. Hori, M. Yoshimura and S. Sōmiya, in Defect Properties and Processing of High-Technology Nonmetallic Materials, edited by Y. Chan, W.D. Kingery and R.J. Stokes (Materials Research Society, Pittsburgh, 1986), pp. 87-94.
9. K. Niihara, Bull. Ceram. Soc. Japan 20, 12 (1985).
10. R.L. Fullman, Trans. AIME 197, 447 (1953).

PROCESSING AND CREEP PERFORMANCE OF SILICON
CARBIDE WHISKER-REINFORCED SILICON NITRIDE

JOHN. R. PORTER*, F. F. LANGE** AND A. H. CHOKSHI***
* Rockwell International Science Center, 1049 Camino Dos Rios, P.O. Box
1085, Thousand Oaks, CA 91360
** Materials Program, University of California, Santa Barbara, CA 93106.
*** Department of Mechanical Engineering, University of California, Davis,
CA 95616.

ABSTRACT

Silicon carbide whisker reinforcement can significantly reduce creep
rates in polycrystalline alumina [1], but the system SiC/Al_2O_3 is thermo-
dynamically unstable in air and oxidizes to mullite during creep testing
[2].* The system SiC/Si_3N_4 was investigated as a potentially more stable,
high temperature structural composite. Silicon carbide whiskers were successfully incorporated into a silicon nitride matrix doped with alumina and
yttria. Processing involved mixing dispersed slurries of silicon carbide
and silicon nitride, adding the dopants as a solution of their nitrates and
subsequently increasing the pH to precipitate the additive hydroxides. The
resulting slurries were filter pressed at room temperature and hot pressed
at 1650°C in graphite dies to full density. X-ray diffraction and transmission electron microscopy confirmed the presence of $\beta-Si_3N_4$, $\alpha-SiC$ and
trace quantities of $\alpha-Si_3N_4$, confirming that the $\alpha-\beta$ Si_3N_4 reaction occurred. An additional, as yet unidentified, minor phase was also detected.
Whisker reinforcement was shown to increase the room temperature
flexural strength and fracture toughness but high temperature creep
performance was unaffected by whisker reinforcement.

INTRODUCTION

The long-term reliability of high temperature structural ceramics is
limited by creep, creep rupture, thermal shock and environmental degradation. A previous investigation demonstrated that the creep resistance of
alumina is significantly improved by SiC whisker reinforcement [1]. The
addition of 15% silicon carbide whiskers to Al_2O_3 reduced creep rates by up
to two orders of magnitude and changed the creep stress exponent from 2 to
5, suggesting a change in creep mechanism [1]. However, conventionally
processed composites were susceptible to oxidation during creep testing and
creep strains-to-failure were small. Failure originated at processing
flaws which, in the conventionally processed composites, were whisker
"nests" and matrix powder hard-agglomerates [2, 3]. Such processing flaws
in ceramic matrix composites can be eliminated through the use of dispersion processing techniques. This approach resulted in increased strains-
to-failure from < 2% outer fiber strain to > 4% during creep testing.

---

*   However, we have found more recently that dispersion-processed alumina
     matrix composites oxidize significantly more slowly than the conven-
     tionally processed composites described in Ref. 2.

EXPERIMENTAL PROCEDURE AND RESULTS

Processing

The silicon nitride matrix composites were fabricated from high purity, predominantly alpha silicon nitride powder, silicon carbide whiskers and sintering additives of yttria and alumina. The silicon carbide whiskers‡ were typically < 1 μm in diameter and up to 80 μm long and were predominantly of alpha form (hexagonal structure). Minor phases in the whisker powder were not detected by x-ray diffraction. The whiskers tend to have a rounded hexagonal cross section with a relatively smooth surface finish. The silicon nitride‡‡ was a submicron powder.

The processing flaws observed in conventionally processed composites are primarily "nests" of whiskers that are devoid of matrix phase and matrix powder "prior-hard-agglomerates" which, after processing, are devoid of whiskers. Additional residual contamination in the as-received whiskers are also potential failure origins in the composites. Sedimentation and flotation techniques were therefore used to separate the whisker nests, other whisker contamination and the matrix agglomerates from their associated slurries [2,3]. Subsequent processing was performed in the slurry state to avoid any further agglomeration.

To prepare the silicon nitride, sedimentation of a dilute dispersion (3% by weight) in water at pH 11 (controlled by additions of tetraethyl ammonium hydroxide) was used to preclude all particulates > 1 μm. The dispersed slurry contained only submicron powder which was then flocculated at pH 8 (controlled by additions of nitric acid). After removing the supernatant water, the slurry was redispersed at pH 11. The silicon carbide whiskers were prepared similarly; 5% by weight of whiskers in water at pH 11 was dispersed ultrasonically and allowed to settle for 10 min. A 1 cm depth of the dispersion which contained floating contamination was discarded from the top. A subsequent 15 cm depth of dispersion was removed and flocculated at pH 2, the supernatant water was removed and the whiskers were redispersed by raising the pH to 11. The weight fraction of whiskers in the slurry after redispersion was typically 10%.

The separate silicon nitride and silicon carbide whisker slurries were then mixed in the correct proportions mechanically and ultrasonically. A 2 M aluminum and yttrium nitrate solution (5:3 molar ratio) was added to the dispersion, which lowered the pH to 5. Tetraethyl ammonium hydroxide was added to cause precipitation of aluminum and yttrium hydroxide until the pH rose to 10. The slurry was then pressure filtered at pressures up to 5 MPa. All pressure filtered bodies were dried at 50°C for 24 h in air, then uniaxially vacuum hot pressed in close fitting graphite dies. Full density was obtained at 1650°C and 24 MPa for 2 h.

Optical observation of polished cross sections indicated that the whiskers were uniformly distributed and that there was a preferred whisker orientation normal to the hot pressing direction. Three different fabricated composites have been characterized in the present investigation. The sample components and their as-fabricated grain size are listed in Table 1. The grain size was defined as 1.74 $\bar{L}$, where $\bar{L}$ is the mean linear intercept grain size, and measurements were taken from polished and etched sample cross sections. The grain size was not uniform in any of the samples, and the measurements included all phases.

X-ray diffraction revealed only trace amounts of α-Si_3N_4 in all samples after hot pressing. Significant Si_2N_2O peaks were present in sample SC 18 (fabricated without SiC) and minor amounts were present in sample

---

‡　ARCO Chemical Co.
‡‡　Ube Corporation

Table I
Silicon Nitride Matrix Composites

Sample ID	v/o SiC	Mole% Additive	Grain Size (μm)
SC 15	15	15	0.84 ± 0.14
SC 17	15	7	0.65 ± 0.10
SC 18	0	0	0.83 ± 0.13

SC 15. All samples contained an as-yet unidentified crystalline phase, which exhibited peaks inconsistent with any known phase in the Si, Al, Y, O, N phase diagrams. However, only trace amounts of this phase were present in sample SC 18.

Room Temperature Mechanical Properties

The room temperature flexural strengths were measured in four-point flexure, using beams 2×3 mm or 3×4 mm in cross section, inner and outer loading spans of 6.5 mm and 29.5 mm and a loading rate of 20 MPa s^{-1}. The indentation fracture toughness was obtained from crack length measurements using a 10 kg indenter load [4]. The indenter was loaded on a polished surface normal to the hot pressing direction so that the crack planes were normal to the plane of preferred whisker orientation. The results are listed in Table II.

Table II
Silicon Nitride Matrix Composite Strength and
Toughness as Function of Whisker Loading

Sample SC	v/o SiC	Flexural Strength (MPa)	Fracture Toughness (MPa m$^{0.5}$)
SC 18	0	850 ± 100	4.2
SC 17	15	1250 ± 100	4.8* 6.0**

* Crack plane parallel to plane of
   predominant whisker alignment.
** Crack plane perpendicular to plane
   of predominant whisker alignment.

Examination of fracture surfaces of failed specimens established that failure typically originated at specimen corners. Regions of the fracture surface close to the failure origins were examined by scanning electron microscopy. Both whiskers lying close to the crack plane and semicylindrical channels from which whiskers had separated can be identified in Fig. 1a. Since the whiskers tend to be randomly distributed normal to the hot pressing direction, few whiskers were oriented suitably for a whisker pull-out toughening mechanism to operate. Typically, whisker failure occurs within 1 μm of the predominant crack plane. Figure 1b shows a number of failed whiskers oriented approximately normal to the crack plane and evidence of significant whisker pull out was not observed. Failure of whiskers oriented almost parallel to the crack plane occurs by bending and the whiskers fail across a whisker diameter.

(a)   **(b)**

Fig. 1. Fracture surface of composite close to failure origin: (a) shows
evidence of whisker separation, (b) shows failed whisker.

Creep Performance

    Creep testing was performed in four-point flexure, using a fixture with
high purity sapphire pivots [5]. The relative displacement of the pivots
was monitored and the applied load and displacement data were converted to
outer fiber stress and strain, using the analysis of Hollenberg et al [6].
    During creep testing in all samples, there was an initial transient of
decreasing creep rate which extended to outer fiber strains of approximately 0.01. During subsequent deformation, creep rates continued to
decrease, albeit at a much lower rate, possibly as a result of concurrent
grain growth. The "steady-state" creep rates reported were taken at the
beginning of the second stage. Figure 2 is a comparison of the creep rates
at 1400°C of a 15 v/o silicon carbide composite with 15 mole% additive
(SC 15), a 15 v/o silicon carbide composite with 7 mole% additive (SC 17),
a monolithic silicon nitride without whisker additions but with 7 mole%
additive (SC 18) and a commercial silicon nitride (Norton Co. NC 132). All
four composites exhibited a creep stress exponent, n, in the creep rela-
tionship $\dot{\varepsilon} = A\sigma^n$, of ≈ 2. There was clearly no reduction in creep rate resulting from whisker addition and, indeed, the most creep resistant mater-
ial was the control monolithic silicon nitride without whisker addition.
    Figure 3 is a comparison of the creep rates of the 15 v/o SiC with
15 mole% additive at 1400 and 1500°C. A similar stress exponent of n ≈ 2
is observed at each temperature.

DISCUSSION

    Whisker reinforcement affects the creep properties of alumina and sili-
con nitride differently. In the whisker reinforced alumina system, the
composite exhibits a higher creep stress exponent than the monolithic
alumina, which suggests that the creep mechanism is altered by the presence
of the whiskers. A creep stress exponent, n, of 2, in a system without a
significant intergranular phase, typically suggests that diffusion creep,
limited by point defect creation/annihilation, is rate controlling, whereas
a stress exponent of 5 is typically attributed to a dislocation creep
mechanism. The reason for the change in mechanism is not understood, but,
presumably, whiskers lying across a grain boundary could inhibit grain
boundary sliding.

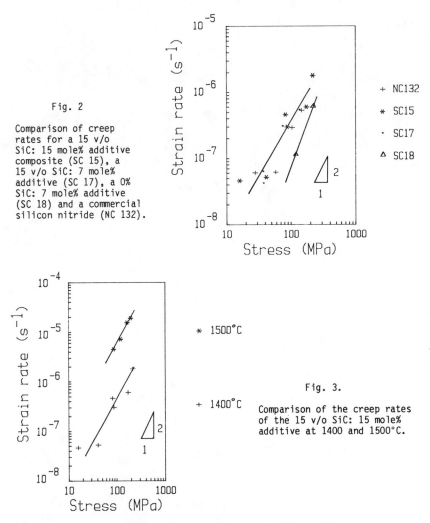

Fig. 2

Comparison of creep rates for a 15 v/o SiC: 15 mole% additive composite (SC 15), a 15 v/o SiC: 7 mole% additive (SC 17), a 0% SiC: 7 mole% additive (SC 18) and a commercial silicon nitride (NC 132).

+ NC132

* SC15

· SC17

△ SC18

* 1500°C

+ 1400°C

Fig. 3.

Comparison of the creep rates of the 15 v/o SiC: 15 mole% additive at 1400 and 1500°C.

However, in the silicon nitride matrix composite, there is no change in the creep stress exponent. This suggests that the effect of the silicon carbide whiskers may not be just a mechanical one. One significant difference between the two systems is the larger amount intergranular phase in the silicon nitride composite. The increased creep rate in the whisker reinforced Si_3N_4 could result from the combined effects of several microstructural variables: (1) Si_3N_4 grain size, (2) impurities from the whiskers reducing the eutectic melting point of the grain boundary phase, (3) the presence of Si_2N_2O and the unidentified phase or phases, and (4) more rapid grain growth in the unreinforced system, with the creep rate being primarily influenced by the Si_3N_4 grain size. Further work will be required to systematically identify the creep-rate limiting microstructural feature. The observation that creep rates continuously decreased with time during creep testing indicates that concurrent grain growth was occurring in these systems.

294

REFERENCES

1. A.H. Chokshi and J.R. Porter, J. Am. Ceram. Soc. 68, C-144 (1985).

2. J.R. Porter and A.H. Chokshi, in Ceramic Microstructures '86: The Role of Interfaces, edited by J. A. Pask and A. G. Evans (Plenum, New York, 1987), in press.

3. J.R. Porter, F.F. Lange and A.H. Chokshi, Ceramic Bull., in press.

4. G.R. Anstis, P. Chantikul, B.R. Lawn and D.B. Marshall, J. Am. Ceram. Soc. 64, 533 (1981).

5. A.H. Chokshi and J.R. Porter, J. Mater. Sci. 21, 705 (1986).

6. G.W. Hollenberg, G.R. Terwilliger, and R.S. Gordon, J. Am. Ceram. Soc. 54, 196 (1971).

DEFORMATION BEHAVIOR OF SiC WHISKER REINFORCED Si$_3$N$_4$

R.D. NIXON,[*] S. CHEVACHAROENKUL,[*] M.L. HUCKABEE,[**] S.T. BULJAN[**] AND R.F. DAVIS[*]
[*]North Carolina State University, Department of Materials Science and Engineering, Box 7907, Raleigh, NC 27695, U.S.A.
[**]GTE Laboratories, Waltham, MA U.S.A.

ABSTRACT

Constant compressive stress creep experiments in the temperature and stress ranges of 1420K - 1570K and 50 MN/m^2 - 350 MN/m^2 have been conducted on hot pressed unreinforced Si$_3$N$_4$ containing the densification aids of 6 wt% Y$_2$O$_3$ and 1.5 wt% Al$_2$O$_3$ and on a similarly prepared composite material reinforced with 20 vol% SiC whiskers. Steady-state creep data obtained on these materials in a N$_2$ atmosphere gave stress exponent values of 1.2 - 1.6 and 0.4 - 1.6 for the unreinforced Si$_3$N$_4$ and the composite, respectively. A break in the steady-state creep rate vs. log stress observed only in the composite occurred at approximately 250 MN/m^2 at the temperatures of 1520K and 1570K. This information coupled with the results of electron microscopy indicate that the temperature as well as the extensive crystallization (or lack of it) of the amorphous grain boundary phase are the principal factors controlling the deformation rate at a given stress within the limits of the small total strains achieved in this research.

INTRODUCTION

Presently available compositions of additive-containing silicon nitride (Si$_3$N$_4$) and silicon nitride solid solutions continue to be candidate materials for a host of structural applications. However, the former has been joined by newly developed ceramic-ceramic composite materials containing SiC whiskers analogous to the Al$_2$O$_3$ and mullite matrix composites introduced and developed by Wei and Becher[1,2]. The employment of unreinforced Si$_3$N$_4$ may be limited for some applications by the fact that its mechanical properties begin to degrade above 1473K as a result of a glass-containing grain boundary phase. This amorphous phase is formed by the reaction between the sintering aid(s) employed to achieve densification and the native SiO$_2$ on the surface of each Si$_3$N$_4$ particle. Whisker reinforcement of this material offers the potential for significant improvements in these properties.

In the present research, constant stress compressive creep experiments have been conducted over a wide range of stresses at three different temperatures on unreinforced and SiC whisker reinforced Si$_3$N$_4$ materials. These studies have been coupled with transmission electron microscopy (TEM) research in order to identify the mechanism(s) responsible for the deformation. The procedures employed and the results obtained from this work as well as the conclusions which may be currently derived are reported in the following sections.

EXPERIMENTAL

The as-received samples were prepared from a mixture of Si$_3$N$_4$[*] derived from vapor phase reaction, 1.5 wt% Al$_2$O$_3$ and 6 wt% Y$_2$O$_3$. The sintering

_____

[*]Type SN-502, GTE Inc., Towanda, PA.

aids of Y_2O_3 and Al_2O_3 produce a more refractory glass upon reaction with the native SiO_2 present on the Si_3N_4 than MgO used in earlier studies. The presence of Al_2O_3 increases the fluidity of the glass relative to that achieved with the use of pure Y_2O_3; as such, the temperature and time necessary to achieve maximum densification are reduced. To produce material for the composites, 20 vol% SiC whiskers[*] in the as-received state were initially dispersed in methanol for 300 s with a sonic probe and subsequently mixed at high speeds with a portion of the Si_3N_4/oxide mixture using a blender. All materials were dried and sieved through 80 and 100 mesh screens. No binders were added at any stage.

The typical and principal impurities in the whiskers have been found [3] by emission spectroscopy to be Al (2810 ppm), Fe (1100 ppm), Ca (2050 ppm), Mn (480 ppm) and Mg (370 ppm). The physical and chemical characteristics of these whiskers have been extensively studied by Nutt [4] and Sharma et al. [5].

All samples were hot pressed for 2-2.5 hrs. at 2000K under a pressure of 34.5 MPa in a flowing N_2 atmosphere at 0.11 MPa total pressure. The power to the hot press was turned off at temperature and the samples allowed to cool at the rate of $\simeq 75°K/min$. to 1273K.

Sections 8 mm in thickness were cut from the original 50 mm dia. x 13 mm thick pressed disks. Cylindrical specimens were subsequently ultrasonically trepanned perpendicular to the hot pressing direction and prepared as described in Refs. [6,7] such that they possessed a height and diameter of 7.6 mm and 3.8 mm, respectively. Each sample was annealed for 4 hrs. at the temperature of creep to allow them to approach a more equilibrium state in the amorphous boundary phase; the samples were not cooled prior to mechanical loading. Very different amounts of crystallization occurred in the boundary phase in the two materials, as described below. Constant stress creep experiments were conducted in continuously purified and circulating N_2 atmosphere at 0.11 MPa pressure using pushrods and secondary platens of W and primary platens of chemically vapor deposited SiC nominally oriented along (111) perpendicular to the applied stress. In this apparatus, uniaxial compressive stresses, constant to within 1% for strains up to 10%, can be applied to the sample. The resulting strains can be read with an accuracy of 0.5 μm. A complete description of this instrument is presented in Ref. [6]. In this research, the creep runs were conducted at constant temperatures within the range of 1420K - 1570K and as a function of stress within the range 50 MN/m² - 350 MN/m² at each temperature. The various conditions for each run are given in Table I. The density of each sample was measured prior to annealing and immediately following each creep experiment. Very slight weight loss was observed after each run, as indicated by the decrease in density for each sample (see Table I).

Following each creep experiment, sections were cut for TEM studies from the center of each crept sample perpendicular to the direction of the applied stress (and parallel to this stress in the reinforced materials), lapped flat, polished to a 1 μm finish, dimpled to a center thickness of

---

[*]ARCO Chemical Co., Advanced Materials Division, Greer, SC. The whiskers were prepared from ground rice hulls which had been mixed with C (in the form of coke) and heated in Ar in an electrical rotary furnace at 1273K to produce a coke-like product containing C and SiO_2. This precursor material was subsequently converted to SiC whiskers in an rf heated graphitizing furnace at 2073K in Ar [3].

Table I. Experimental creep data for the unreinforced and SiC whisker reinforced silicon nitride.

Sample #	Densities, g/cm^3		Anneal/Creep Temps.(K)	Anneal Time (hr)	Stress (MN/m^2)	$\dot{\varepsilon}_{ss}$ (s^{-1})	% Total Strain
	Preanneal	Post Creep					
2	3.238	3.238	1420	4	300	0	0
3	3.239	3.234	1470	4	50	0	
					100	0	
					150	2.20E-9	
					200	2.97E-9	
					250	4.57E-9	
					300	6.36E-9	
					350	7.07E-9	0.22
4	3.237	3.219	1520	4	50	1.37E-9	
					100	5.61E-9	
					150	8.49E-9	
					200	1.11E-8	
					250	1.56E-8	
					300	2.04E-8	
					350	2.69E-8	0.66
5	3.234	3.175	1570	4	50	3.59E-9	
					100	7.35E-9	
					150	1.22E-8	
					200	1.90E-8	
					250	2.30E-8	
					300	3.33E-8	
					350	4.24E-8	1.34
2W*	3.232	3.222	1470	4	50	0	
					100	1.51E-9	
					150	2.98E-9	
					200	5.26E-9	
					250	6.02E-9	
					300	6.95E-9	
					350	9.64E-9	0.33
3W	3.230	3.202	1520	4	50	4.90E-9	
					100	9.30E-9	
					150	1.08E-8	
					200	1.15E-8	
					250	1.39E-8	
					300	1.73E-8	
					350	2.26E-8	0.83
4W	3.230	3.205	1570	4	50	1.80E-8	
					100	2.48E-8	
					150	3.02E-8	
					200	3.64E-8	
					250	4.66E-8	
					300	6.19E-8	1.15

*W denotes SiC whisker reinforced material

25 μm and thinned to electron transparency using the conditions of 6 KV, 1 amp at 12° gun tilt until perforation and 3 KV, 1 amp at 5° for 0.25 hr. and subsequently C-coated to prevent charging during analysis.

RESULTS AND DISCUSSION

The microstructure of both the as-received Si_3N_4-based materials, as observed via TEM, contained virtually dislocation-free grains of β'-Si_3N_4 having an average grain size (determined by mean intercept procedures) of $\approx$ 0.3 μm and surrounded primarily by an amorphous phase. However, occasional crystallization of this amorphous phase was evident. A representative micrograph which illustrates these features for both materials is shown in Fig. 1a. Energy dispersive analysis revealed the presence of Y and Si in the boundary phase; a very small contribution from Al was also detected in both sample types in this phase. These results are similar to those reported by Krivanek et al. [8] and Ahn and Thomas [9]. In contrast to the findings of these investigators, Y was also detected inside the $\beta'-Si_3N_4$ grains.

Annealing of the unreinforced samples at each of the temperatures of creep for 4 hrs. immediately prior to deformation resulted in extensive crystallization of the boundary phase, as shown in Fig. 1b. Although electron diffraction was employed in an attempt to identify this phase, the values of the d spacings could not be matched with a known Si_3N_4 solid solution. The largest d spacing from lattice fringe imaging was 0.695 nm. In addition, this phase was everywhere separated from the $\beta'-Si_3N_4$ by a very thin amorphous phase similar to that frequently found at $\beta'-\beta'$ boundaries. Thus the initial crystallization appears to have been nucleated in the glass phase rather than epitaxially on the $\beta'-Si_3N_4$.

By contrast, annealing of the whisker-containing materials caused an obviously smaller amount of crystallization relative to the unreinforced material, as shown in Fig. 1c. This more stable glass is believed to be produced by the diffusion of the glass modifying elements of Ca, Mn, Fe and Mg from the SiC whisker into the amorphous boundary phase and the consequent movement of Al and perhaps some of the Y into Si (glass former) positions. Indeed the presence of the SiC is the only initial difference between the samples. The density change between the as-received and crept samples is small, as shown in Table I; thus very little weight loss occurred during deformation. Also, the porosity in both samples before and after creep was < 1%. Finally, the length and diameter range of the SiC whiskers in the as-received material were 6-10 μm and 0.2 - 0.5 μm, respectively. Preferred orientation of this phase toward directions perpendicular to the axis of hot pressing was evident in the samples from which the TEM specimens were taken.

Creep in both types of materials at each increment of stress at all temperatures was characterized by a period of primary creep in which the strain rate was high followed by a period of steady-state creep, as shown in the representative curves of Fig. 2. The objective of the research described herein has been to characterize the deformation process within the limit of the total strain of 1.5% (see Table I). At the outset it should be noted that under all conditions employed in this research, both Si_3N_4-based materials showed excellent creep resistance. The values of the stress exponents (from the expression $\dot{\varepsilon}_{ss} \alpha \sigma^n$) as shown in Fig. 3 for the unreinforced material are between 1 and 2 and are indicative of a grain boundary sliding process. These values are constant throughout the range of stresses employed at a given temperature but decrease with increasing temperature as would be expected for this type of process. The steady-state creep rates ($\dot{\varepsilon}_{ss}$) and the stress exponent for the whisker-containing material are very similar to those measured for the unreinforced material at 1470K, as shown in Fig. 4. However, at 1520K and 1570K the values of both parameters in the reinforced material change dramatically at the lower stresses from the analogous values in the unreinforced material. The values of $\dot{\varepsilon}_{ss}$ in the latter material are much higher at all stresses between 50 and ≃ 250 MN/m². However, the stress dependence of $\dot{\varepsilon}_{ss}$ was considerably reduced in the reinforced material in the aforenoted stress range. In addition a break occurs in the two higher temperature curves of the reinforced material at ≃ 250 MN/m² such that the n values and steady-state creep rates become essentially equal to (at 1520K) or larger than (at 1570K) the values for the unreinforced material.

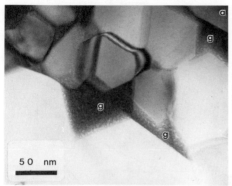

Figure 1. TEM micrographs of a)
the representative microstructure
of both the as—hot pressed
unreinforced and SiC whisker
reinforced Si_3N_4, and of the
former (b) and the latter (c)
materials annealed at 1520K for
four hours. (g = amorphous
phase; C = phase crystallized
from g.)

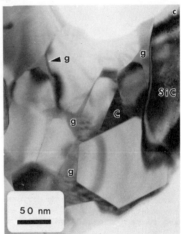

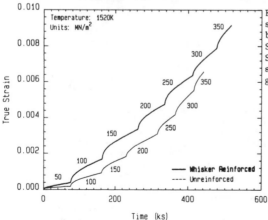

Figure 2. Creep curves
sequentially obtained for
both the unreinforced and
SiC whisker reinforced
Si_3N_4 at 1520K and the
stresses shown on the
graph.

300

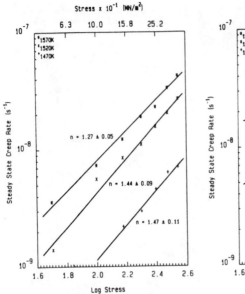

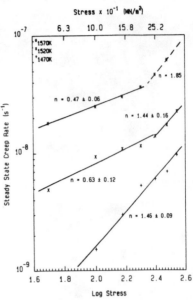

Figure 3. Steady-state creep rate values as a function of log stress using the data from Table I for the unreinforced Si_3N_4.

Figure 4. Steady-state creep rate values as a function of log stress using the data from Table I for the SiC whisker reinforced Si_3N_4

As noted above, extensive crystallization occurred in the grain boundary amorphous phase of the unreinforced material during annealing, in contrast to the composite material which became only partially crystalline. However, the microstructure of all the samples at the conclusion of the creep experiments showed extensive crystallization, as shown in Figure 5A and B. No evidence of relative grain motion or the formation of triple point voids or grain boundary porosity or grain growth was discerned in the samples crept for the small total strains noted in Table I. Occasional dislocation tangles were observed in large untransformed $\alpha\text{-}Si_3N_4$ grains but were not directly associated with the deformation.

A synthesis of the above data indicates that the primary stage of creep observed under all conditions was caused by rather rapid but very small adjustments in the position of the grains in the material. Both elastic and plastic (viscous flow) components are believed to play concurrent roles in this stage of deformation. However in the steady-state region, the data indicates that viscous flow is the dominant mechanism of creep in both types of materials. It is believed that this mechanism is hindered by the presence of the crystalline grain boundary phase, particularly in the unreinforced material crept at 1520K and 1570K. At 1470K, the values of the creep rates at all stresses are very low and virtually identical for both materials indicating that the glass phase is very viscous at this temperature and that the amount of crystallization in the boundary phase exerts relatively limited control on the deformation rate at any stress. Moreover, most of the remaining glass in the composite material crystallized during the creep experiments at 1470K with no effect

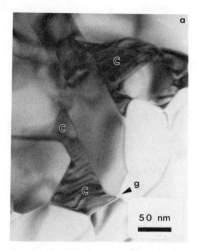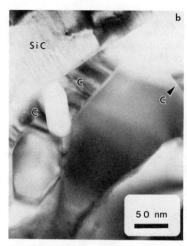

Figure 5. TEM micrographs of A) the unreinforced and B) the SiC whisker reinforced Si_3N_4 crept under the final conditions of 1520K and 350 MN/m^2. (g = amorphous phase; C = phase crystallized from g.)

on the stress exponent. Increasing the temperature to 1520K or 1570K lowers the viscosity of the residual glass for both materials, but particularly for the SiC-containing material because of incorporation of impurities from the whisker into the glass phase. In addition, the smaller amount of crystalline boundary phase in the latter material may also contribute to the marked jump in creep rate relative to the unreinforced material. Finally, the break in the curves of the 1520K and 1570K reinforced material are believed to be caused by the increased crystallization of the glass phase rather than the onset of hindrance to deformation by the whisker phase. The possibility of a stress induced transformation or changes in the chemical composition of the glass in the composite are now being investigated.

Future research will include the determination of (1) the conditions of temperature, stress and time which favor the crystallization of the composite and (2) the reason for the sluggishness of this transformation relative to the unreinforced material. Additional studies regarding the physical and chemical nature of the crystalline boundary phase(s), the activation energies for steady-state creep, the effect of higher total strains and increased whisker loading on deformation rate and whisker-Si_3N_4 interaction will be conducted.

SUMMARY

Constant compressive stress creep experiments over total strains of < 1.5% on unreinforced and SiC whisker reinforced $\beta'-Si_3N_4$ containing crystalline and amorphous boundary phases show that at 1470K both materials deform at the same rates within the stress limits of 50 - 350 MN/m^2. Increasing the temperature to 1520K or 1570K causes a larger increase in values of $\dot{\epsilon}_{ss}$ of the composite relative to the unreinforced material up to 250 MN/m^2; a significant decrease in the stress exponent also occurs in the former material. Above 250 MN/m^2 the stress exponents become equal. All of these phenomena are believed to be affected by the extent of the crystallization of the amorphous boundary phase, the temperature and the viscosity of the residual glass phase. The small amount of total strain does not allow the SiC whiskers per se to play any direct role in controlling the deformation.

ACKNOWLEDGEMENTS

The authors acknowledge support of the Army Research Office (DAAL03-86-K-0013) for support of this project and to C.H. Carter, Jr. and S.R. Nutt for helpful discussions. The whisker-toughened material used in this study was developed at GTE Laboratories through research sponsored by the U.S. Department of Energy, as part of the Ceramic Technology for Advanced Heat Engine Project under contract DE-AC05-840R21400 with Martin Marietta Energy Systems, Inc. (Oak Ridge National Laboratory).

REFERENCES

1. P.F. Becher, G.C. Wei, J. Am. Ceram. Soc. 67, C267 (1984).
2. G.C. Wei and P.F. Becher, Am. Ceram. Soc. Bull. 64, 298 (1985).
3. S.R. Nutt, Private communication.
4. S.R. Nutt, J. Am. Ceram. Soc. 67, 428 (1984).
5. N.K. Sharma, W.S. Williams and A. Zangvil, J. Am. Ceram. Soc. 67, 715 (1984).
6. C.H. Carter, Jr., C.A. Stone, R.F. Davis and D.R. Schaub, Rev. Sci. Instrum. 51, 1352 (1980).
7. C.H. Carter, Jr., and R.F. Davis, J. Am. Ceram. Soc. 67, 409 (1984).
8. O.L. Krivanek, T.M. Shaw and G. Thomas, J. Am. Ceram. Soc. 62, 585 (1979).
9. C.C. Ahn and G. Thomas, J. Am. Ceram. Soc. 16, 14 (1983).

Author Index

Subject Index

306